# BOUNDARY VALUE PROBLEMS FOR HIGHER ORDER DIFFERENTIAL EQUATIONS

# BOUNDARY VALUE PROBLEMS FOR HIGHER ORDER DIFFERENTIAL EQUATIONS

## Ravi P. Agarwal

Department of Mathematics
National University of Singapore, Singapore

**World Scientific**

*Published by*

**World Scientific Publishing Co Pte Ltd.**
**P. O. Box 128, Farrer Road, Singapore 9128**
242, Cherry Street, Philadelphia PA 19106-1906, USA

Library of Congress Cataloging-in-Publication Data

Agarwal, Ravi P.
  Boundary value problems for higher order differential equations.

  Includes index.
  1. Boundary value problems.
QA379.A34          1986          515.3'5          86-11016
ISBN 9971-50-108-2

Printed in Singapore by Fu Loong Lithographer Pte Ltd.

In memory of my Sister, Manorama Agarwal

# PREFACE

This monograph, an in-depth and up-to-date coverage of more than 250 recent research publications on boundary value problems for ordinary differential equations, aims at the needs of graduate students, numerical analysts as well as researchers who are looking for open problems.

The first draft of this monograph was written at der Ludwig - Maximilians - Universität, Munich where I spent two years as an Alexander von Humboldt Foundation Fellow. It is my pleasant duty to place on record my sincere tribute to the Foundation. Thanks are also due to Professor G. Hämmerlin for his inspiration to embark on this project. My thanks are also due to Professors R. Conti, L. Jackson and V. Lakshmikantham for their continued inspiration and guidance. I am also grateful to various colleagues, friends and students who have helped me at various stages during the preparation of the manuscript. Finally my appreciation and thanks go to Mdm Tay Lee Lang for her excellent and careful typing.


Singapore                                          Ravi. P. Agarwal
April, 1986

# CONTENTS

x

# BOUNDARY VALUE PROBLEMS
# FOR HIGHER ORDER
# DIFFERENTIAL EQUATIONS

# 1. SOME EXAMPLES

Boundary value problems (hereafter to be abbreviated as BVPs) manifest themselves in almost all branches of Science, Engineering and Technology, for instance, boundary layer theory in fluid mechanics, heat power transmission theory, space technology and also control and optimization theory, to cite only a few. The following examples provide a variety of situations of occurrence of BVPs and the motivation for some of the problems we shall consider in these notes.

Example 1.1    In the problem of the motion of a particle of mass m under the action of a given force $\bar{F}(t, \bar{r}, \bar{r}')$ it is frequently necessary to find the law of motion if at the initial time $t = t_0$ the particle was located in a position characterized by the radius vector $\bar{r}_0$ and at time $t = t_1$ it has reached the point $\bar{r} = \bar{r}_1$. The problem reduces to integrating the differential equation of motion

$$m \frac{d^2\bar{r}}{dt^2} = \bar{F}(t, \bar{r}, \bar{r}')$$

with the boundary conditions $\bar{r}(t_0) = \bar{r}_0$, $\bar{r}(t_1) = \bar{r}_1$ .

Example 1.2    A natural and prolific source of nonlinear differential equations with two point boundary conditions is the calculus of variations. Consider the problem of finding the extrema of the functional

$$\int_a^b F(t, x(t), x'(t))dt$$

under the conditions $x(a) = \alpha$, $x(b) = \beta$. Suitable hypotheses on F lead to a second order BVP (Euler's equation $F_x - \frac{d}{dt} F_{x'} = 0$ with $x(a) = \alpha$, $x(b) = \beta$).

2

Example 1.3    We consider similarity solution of the unsteady flow of
gas through a semi-infinite porous medium, initially filled with gas
at a uniform pressure $P_0$. At time $t = 0$ the pressure at the out-flow
face is suddenly reduced from $P_0$ to $P_1$ and thereafter maintained at
this lower pressure.   In terms of a dimension free quantity w, defined
by

$$w(z) = \alpha^{-1}(1 - \frac{P^2(z)}{P_0^2}), \quad \alpha = 1 - \frac{P_1^2}{P_0^2}$$

the problem takes the form

$$w''(z) + \frac{2z}{\sqrt{1 - \alpha w(z)}} \, w'(z) = 0; \quad w(0) = 1, \; w(+\infty) = 0.$$

Example 1.4    In the case of BVPs a small change in the boundary conditions can lead to significant changes in the behavior of the solution.

Let us consider the initial value problem $x''' + x = 0$; $x(0) = c_1$,
$x'(0) = c_2$, $x''(0) = c_3$.   It has the unique solution

$$x(t) = \frac{1}{3}(c_1 - c_2 + c_3)e^{-t} + \frac{1}{3}(2c_1 + c_2 - c_3)e^{t/2} \cos \frac{\sqrt{3}}{2}t + \frac{1}{\sqrt{3}}(c_2 + c_3)e^{t/2} \sin \frac{\sqrt{3}}{2}t$$

for any set of values $c_1$, $c_2$, $c_3$.

However, the BVP  $x''' + x = 0$; $x(0) = 0$, $x'(0) = 0$, $x(b_1) = \varepsilon (\neq 0)$
where $b_1$ is the first positive root of the equation $2 \sin(\frac{\sqrt{3}}{2}b - \frac{\pi}{6})$
$+ e^{-3/2b} = 0$ has no solution; the problem $x''' + x = 0$; $x(0) = 0$, $x'(0) = 0$,
$x(b) = \varepsilon$, $0 < b < b_1$ has the unique solution

$$x(t) = \frac{\varepsilon \exp(\frac{1}{2}(t-b))[\exp(-\frac{3}{2}t) + 2 \sin(\frac{\sqrt{3}}{2}t - \frac{\pi}{6})]}{[2 \sin(\frac{\sqrt{3}}{2}b - \frac{\pi}{6}) + \exp(-\frac{3}{2}b)]}$$

while the problem $x''' + x = 0$; $x(0) = 0$, $x'(0) = 0$, $x(b_1) = 0$ has an infinite number of solutions

$$x(t) = k[e^{-t} + 2e^{t/2} \sin(\frac{\sqrt{3}}{2} t - \frac{\pi}{6})]$$

where k is an arbitrary constant.

This example naturally raises the following fundamental questions in regard to a given BVP :

$(P_1)$    Does a solution exist ?

$(P_2)$    If so, is that solution unique ?

$(P_3)$    If the solution is unique, what is the maximum length of the interval in which uniqueness holds ?

Example 1.5    The BVP

(1.1)                      $x'' = \lambda e^{\alpha x}$, $x(0) = x(1) = 0$

arises in applications involving the diffusion of heat generated by positive temperature-dependent sources.  For instance, if $\alpha = 1$ it arises in the analysis of Joule losses in electrically conducting solids, with $\lambda$ representing the square of the constant current and $e^x$ the temperature-dependent resistance, or in frictional heating with $\lambda$ representing the square of the constant shear stress and $e^x$ the temperature-dependent fluidity.

If $\alpha\lambda = 0$, the problem (1.1) has a unique solution

(i)      if $\lambda = 0$ then, $x(t) \equiv 0$

(ii)     if $\alpha = 0$ then, $x(t) = \frac{\lambda}{2} t(t-1)$.

If $\alpha\lambda < 0$, the problem (1.1) has as many solutions as the number of roots of the equation $c = \sqrt{2|\alpha\lambda|} \cosh\frac{c}{4}$, also for each such $c_i$ the solution is

$$x_i(t) = -\frac{2}{\alpha}\{\log[\cosh(\frac{c_i}{2}(t-\frac{1}{2}))] - \log[\cosh(\frac{c_i}{4})]\}.$$

From the equation $c = \sqrt{2|\alpha\lambda|}\cosh\frac{c}{4}$, it follows that if

$$\sqrt{|\alpha\lambda|/8}\ \min_{c\geq 0}\frac{\cosh c/4}{c/4} \quad \begin{cases} < 1, & (1.1) \text{ has two solutions} \\ = 1, & (1.1) \text{ has one solution} \\ > 1, & (1.1) \text{ has no solution.} \end{cases}$$

If $\alpha\lambda > 0$, the problem (1.1) has as many solutions as the number of roots of the equation $c = \sqrt{2\alpha\lambda}\cos\frac{c}{4}$, also for each such $c_i$ the solution is

$$x_i(t) = \frac{2}{\alpha}\log[c_i/\cos(\frac{c_i}{2}(t-\frac{1}{2}))] - \frac{1}{\alpha}\log[2\alpha\lambda].$$

Form the equation $c = \sqrt{2\alpha\lambda}\cos\frac{c}{4}$, it follows that if

$$\sqrt{\alpha\lambda/8}\quad \max_{\frac{3\pi}{4}\leq\frac{c}{4}\leq\frac{5\pi}{2}}\frac{\cos c/4}{c/4} \quad \begin{cases} < 1, & (1.1) \text{ has one solution} \\ \geq 1, & (1.1) \text{ has at least two solutions.} \end{cases}$$

Thus, in particular if $\lambda = 1$ and $\alpha = -1$ the BVP (1.1) has two solutions $x_1(t)$ and $x_2(t)$. Solution $x_1(t)$ drops below upto $-0.14050941...$ and $x_2(t)$ upto $-4.0916146...$. Hence, if we restrict the region under consideration to $[0,1] \times \{x: |x| < 4.0916146...\}$ then, there is a unique solution. From this example we are led to ask the additional question :

($P_4$)   What is the greatest region in which a given BVP can have a unique solution ?

Example 1.6   Beams formed by a few lamina of different materials are known as sandwich beams.  In an analysis of such beams subject to uniformly distributed load along the entire length, Krajcinvic [11] found that the distribution of shear deformation x is governed by the linear differential equation

(1.2)   $$x''' - k^2 x' + a = 0$$

where $k^2$ and a are physical constants which depend on the elastic properties of the lamina.

For the free ends, the condition of zero shear bimoment at both ends leads to the boundary conditions

(1.3)   $$x'(0) = x'(1) = 0.$$

From symmetry considerations

(1.4)   $$x(\tfrac{1}{2}) = 0.$$

The three point BVP (1.2)-(1.4) has a unique solution and can be expressed in terms of elementary functions

$$x(t) = \frac{a}{k^3}[(\sinh \tfrac{1}{2} k - \sinh kt) + k(t - \tfrac{1}{2}) + \tanh \tfrac{1}{2} k(\cosh kt - \cosh \tfrac{1}{2} k)]$$

however, in general even if it is known that the given BVP has a unique solution it is not possible to find it explicitly.  Thus, we must seek :

(P$_5$)   Constructive methods for a given BVP.

Example 1.7   Imposing some ideal conditions the deflection of the beam analysis leads to a fourth order linear differential equation

$$(p(t)x'')'' + q(t)x = r(t)$$

together with one of the following sets of the boundary conditions

(i)  $x(a) = A_1$, $x''(a) = A_2$, $x(b) = B_1$, $x''(b) = B_2$

(ii)  $x(a) = A_1$, $x'(a) = A_2$, $x(b) = B_1$, $x'(b) = B_2$

(iii)  $x(a) = A_1$, $x'(a) = A_2$, $x''(b) = B_1$, $x'''(b) = B_2$

(iv)  $x''(a) = A_1$, $x'''(a) = A_2$, $x''(b) = B_1$, $x'''(b) = B_2$.

In particular the transverse displacement of an elastically im-
bedded rail to a distributed transverse load is described by the linear
fourth order equation

(1.5)  $$(EJ(\xi)u'')'' + K(\xi)u = g(\xi)$$

where $EJ(\xi)$ is the flexural rigidity, $K(\xi)$ the elastic resistance of
the supporting material and $g(\xi)$ the load density. For a freely sup-
ported rail the boundary conditions are

$$u''(-L) = u'''(-L) = u''(L) = u'''(L) = 0$$

and correspond to vanishing moments and shear forces at the rail ends.
When EJ, g and K are given by $EJ(\xi) = EJ_0(2 - (\xi/L)^2)$, $g(\xi) = (2 - (\xi/L)^2)$,
the equation (1.5) can be converted into the nondimensional form

(1.6)  $$((2 - t^2)x'')'' + 40x = 2 - t^2$$

where $t = \xi/L$, $x = (\dfrac{EJ_0}{L^4})u$, $(\dfrac{L^4}{EJ_0})\, K = 40$ (we assume).

Example 1.8  Although we shall be dealing only with higher order
differential equations, for computational purpose sometimes it

becomes necessary to convert the given BVP into an equivalent system. For example, let $y = (2 - t^2)x''$, then the equation (1.6) is equivalent to the second order system

$$(2 - t^2)x'' - y = 0$$

$$y'' + 40x = 2 - t^2.$$

Further, it we take $u = (x, x', y, y')^T$ then, the above system is equivalent to the first order system

$$(1.7) \quad \begin{vmatrix} u_1' \\ u_2' \\ u_3' \\ u_4' \end{vmatrix} = \begin{vmatrix} 0 & 1 & 0 & 0 \\ 0 & 0 & 1/(2-t^2) & 0 \\ 0 & 0 & 0 & 1 \\ -40 & 0 & 0 & 0 \end{vmatrix} \begin{vmatrix} u_1 \\ u_2 \\ u_3 \\ u_4 \end{vmatrix} + \begin{vmatrix} 0 \\ 0 \\ 0 \\ 2-t^2 \end{vmatrix} .$$

We shall assume that the rail is hinged in a complicated manner and its end points and that the moments and shear forces have to be determined at the rail ends from the measured displacements at four different points along the rail. For this, let the following dimensionless displacements be observed

$$u_1(0.2) = 0.0448156$$

$$u_1(0.4) = 0.0433224$$

$$(1.8)$$

$$u_1(0.6) = 0.0410152$$

$$u_1(0.8) = 0.0381534.$$

System (1.7) subject to the boundary conditions (1.8) constitutes a four point BVP. Obviously, it is impossible to find the solution of (1.7), (1.8) in terms of elementary functions.

# COMMENTS AND BIBLIOGRAPHY

Example 1.3 is an infinite interval problem and first appeared in [10], also see [6, 13]. Example 1.4 is given in [1]. Example 1.5 is taken from [2] and for several particular cases see [8, 9, 14, 15]. A very general discussion of Sandwich beam analysis has been given recently in [5]. Example 1.7 is of paramount interest in engineering and in the last ten years Usmani [16-20] has examined several finite difference methods to construct the solution, also see [3, 4, 7]. Example 1.8 is taken from [12].

1.  Agarwal, R. P. "Nonlinear two-point boundary value problems", Proc. Seventeenth Conf. Theo. Appl. Mech. (1972).
2.  Agarwal, R. P. and Loi, S. L. "On approximate Picard's iterates for multipoint boundary value problems", Nonlinear Analysis : Theory, Methods and Appl. 8, 381-391 (1984).
3.  Agarwal, R. P. and Chow, Y. M. "Iterative methods for a fourth order boundary value problem", J. Comp. Appl. Math. 10, 203-217 (1984).
4.  Agarwal, R. P. and Wilson, S. J. "On a fourth order boundary value problem", Utilitas Mathematica 26, 297-310 (1984).
5.  Agarwal, R. P. "Existence-uniqueness and iterative methods for third order boundary value problems",J. Comp. Appl. Math. to appear.
6.  Bailey, P. B., Shampine, L. F. and Waltman, P. E. Nonlinear Two Point Boundary Value Problems, Academic Press, New York, 1968.
7.  Collatz, L. The Numerical Treatment of Differential Equations, Springer-Verlag, New York, 1966.
8.  Kalaba, R. "On nonlinear differential equations, the maximum operation, and monotone convergence", J. Math. Mech. 8, 519-574 (1959).
9.  Keller, J. "Electrohydrodynamics I : The equilibrium of a charged gas in a container", J. Rat. Mech. Anal. 5, 715-724 (1956).
10. Kidder, R. E. "Unsteady flow of gas through a semi-infinite porous medium", J. Appl. Mech. 27, 329-332 (1957).
11. Krajcinvic, D. "Sandwich beam analysis", J. Appl. Mech. 39, 773-778 (1972).
12. Meyer, G. H. Initial Value Methods for Boundary Value Problems, Theory and Application of Invariant Imbedding, Academic Press, New York, 1973.
13. Na, T. Y. Computational Methods in Engineering Boundary Value Problems, Academic Press, New York, 1979.
14. Shampine, L. F. "Monotone iterations and two-sided convergence", SIAM J. Numer. Anal. 3, 607-615 (1966).

15. Talbot, Th. D. "Guaranteed error bounds for computed solutions of nonlinear two-point boundary value problems",MRC Technical Summary Report, 875, University of Wisconsin, Madison, 1968.

16. Usmani, R. A. "On the numerical integration of a boundary value problem involving a fourth order linear differential equation", BIT 17, 227-234 (1977).

17. Usmani, R. A. "Discrete variable methods for a boundary value problem with engineering applications", Mathematics of Computation 32, 1087-1096 (1978).

18. Usmani, R. A. "Discrete methods for boundary value problems with applications in plate deflection theory", Jour. Appl. Math. Phys. 30, 87-99 (1979).

19. Usmani, R. A. "Solving boundary value problems in plate deflection theory", Simulation, 195-206, December 1981.

20. Usmani, R. A. and Taylor, P. J. "Finite difference methods for solving [p(x)y"]" + q(x)y = r(x)", Intern. J. Computer Math. 14, 277-293 (1983).

## 2. LINEAR PROBLEMS

For the nth order linear differential equation

$$(2.1) \qquad L[x] = x^{(n)} + \sum_{i=1}^{n} p_i(t) x^{(n-i)} = f(t)$$

where $p_i(t)$, $1 \le i \le n$ and $f(t)$ are continuous on $I = [a,b]$, we shall consider the following separated boundary conditions

$$(2.2) \qquad \ell_i[x] = \sum_{k=0}^{n-1} c_{ik} x^{(k)}(a_i) = A_i, \quad 1 \le i \le n$$

where $a \le a_1 \le a_2 \le \cdots \le a_n \le b$.

In (2.2) the coincidence of several $a_i$ means that at a single point several functions are defined, which are assumed to be linearly independent. These boundary conditions include in particular the

(i)   Initial conditions

$$a_1 = a_2 = \cdots = a_n$$

(ii)   Nicoletti conditions

$$a_1 < a_2 < \cdots < a_n$$

$$(2.3) \qquad x(a_i) = A_i \,; \quad i = 1, 2, \ldots, n$$

(iii) Hermite (r point) conditions

$$a_1 < a_2 < \cdots < a_r \ (r \ge 2), \ 0 \le k_i, \ \sum_{i=1}^{r} k_i + r = n$$

$$(2.4) \qquad x(a_i) = A_{1,i}, \ x'(a_i) = A_{2,i}, \ldots, x^{(k_i)}(a_i) = A_{k_i+1,i};$$

$$i = 1, 2, \ldots, r$$

(iv)  Abel-Gontscharoff conditions

$$a_1 \le a_2 \le \cdots \le a_n \ (a_n > a_1)$$

(2.5)     $x^{(i)}(a_{i+1}) = A_i \ ; \quad i = 0,1,\ldots,n-1$

and in particular the right focal point boundary conditions

$$a_1 < a_2, \ 1 \le k \le n-1 \text{ is fixed}$$

$$x^{(i)}(a_1) = A_i \ ; \quad i = 0,1,\ldots,k-1$$

(2.6)

$$x^{(i)}(a_2) = A_i \ ; \quad i = k,k+1,\ldots,n-1$$

(v)   (n-1,p) conditions

$$a_1 < a_2, \ 0 \le p \le n-1 \text{ is fixed}$$

$$x^{(i)}(a_1) = A_i \ ; \quad i = 0,1,\ldots,n-2$$

(2.7)

$$x^{(p)}(a_2) = B$$

(vi)  (p,n-1) conditions

$$a_1 < a_2, \ 0 \le p \le n-1 \text{ is fixed}$$

$$x^{(p)}(a_1) = B$$

(2.8)

$$x^{(i)}(a_2) = A_i \ ; \quad i = 0,1,\ldots,n-2$$

(vii)  Lidstone conditions

$$a_1 < a_2, \ n = 2m$$

(2.9)     $x^{(2i)}(a_1) = A_{2i}, \ x^{(2i)}(a_2) = B_{2i} \ ; \quad i = 0,1,\ldots,m-1.$

Since the interpolating polynomial satisfying (2.2) is a solution of the differential equation $x^{(n)} = 0$, the BVP (2.1), (2.2) gives the possibility of interpolation by the solutions of the differential equation (2.1).

To discuss the existence and uniqueness of the solutions of the BVP (2.1), (2.2) we need the following :

Lemma 2.1 [10]   Consider the system of n linear equations

(2.10) $$Ax = b$$

where A is an n × n matrix and x and b are n dimensional vectors. Then, if

(i)   rank A = n

the system (2.10) possesses a unique solution. Alternatively, the homogeneous system

$$Ax = 0$$

possesses only the trivial solution.

(ii)   rank A = n - m (1 $\leq$ m $\leq$ n)

the system (2.10) possesses a solution if and only if

(2.11) $$\Delta b = 0$$

where $\Delta$ is an m × n matrix whose row vectors are linearly independent vectors $d_\alpha$ , 1 $\leq$ $\alpha$ $\leq$ m satisfying

$$d_\alpha A = 0.$$

In case (2.11) holds, any solution of (2.10) can be given by

$$x = \sum_{\alpha=1}^{m} k_\alpha c_\alpha + Sb$$

where $k_\alpha$ , $1 \leq \alpha \leq m$ are arbitrary constants, $c_\alpha$, $1 \leq \alpha \leq m$ are m
linearly independent column vectors satisfying

$$Ac_\alpha = 0$$

and S is an $n \times n$ matrix independent of b such that

$$ASp = p$$

for any column vector p satisfying

$$\Delta p = 0.$$

Let $\{x_j(t)\}$, $1 \leq j \leq n$ be any fundamental system of solutions of
the homogeneous differential equation

(2.12) $$L[x] = 0$$

and $\phi(t)$ be any particular solution of (2.1). Then, any solution of
(2.1) can be written as

(2.13) $$x(t) = \sum_{j=1}^{n} \alpha_j x_j(t) + \phi(t)$$

where $\alpha_j$, $1 \leq j \leq n$ are unknown constants.

The solution (2.13) satisfies the boundary conditions (2.2) if and
only if

(2.14) $$A_i = \ell_i[\sum_{j=1}^{n} \alpha_j x_j + \phi] = \sum_{j=1}^{n} \ell_i[x_j]\alpha_j + \ell_i[\phi], \quad 1 \leq i \leq n.$$

From Lemma 2.1 the system (2.14) has a unique solution if and only
if

(2.15) $$\det(\ell_i[x_j]) \neq 0, \quad 1 \leq i, j \leq n.$$

Thus, the existence and uniqueness of the solution of (2.1), (2.2) is equivalent to the corresponding homogeneous problem : (2.12) together with

(2.16) $$\ell_i[x] = 0, \quad 1 \le i \le n$$

has no nontrivial solution.

Further, from Lemma 2.1 we note that if the rank of the matrix $(\ell_i[x_j])$ is $n - m$ $(1 \le m \le n)$ then also (2.1), (2.2) may have a solution. However, this solution will contain m arbitrary constants, i.e., the uniqueness is lost.

Hence, we see that the uniqueness of the solution of (2.1), (2.2) implies the existence, whereas the lack of the uniqueness does not decide the existence. This uniqueness implies existence property of linear BVPs is not so immediate for the nonlinear BVPs and we need to answer :

($P_6$)  Does the uniqueness of a given nonlinear BVP imply existence ?

## COMMENTS AND BIBLIOGRAPHY

Nicoletti [8] considered the BVP (2.1), (2.3) as early as the year 1897. Obviously, the boundary conditions (2.4) include (2.3) as a particular case. For the early work on the BVP (2.1), (2.4) see [4, 9, 11, 12], whereas the recent work is contained in numerous number of papers some of which we shall refer to in later sections. When we consider a problem of applied sciences or engineering described by (2.1), we are unable to measure all $x^{(i)}(t)$, $0 \le i \le n-1$ at the same instant $a_1$ of time because our instruments have limited accuracy. Therefore, the BVP (2.1), (2.5) has practical importance and some of its particular cases have been discussed in [3, 7]. The BVPs (2.1), (2.7) and (2.1),

(2.8) arise in determining intervals of nonoscillation [5, 6]. The BVP (2.1), (2.9) has been considered in [1, 2].

1.  Agarwal, R. P. and Akrivis, G. "Boundary value problems occurring in plate deflection theory", J. Comp. Appl. Math. 8, 145-154 (1982).
2.  Chawla, M. M. and Katti, C. P. "Finite difference methods for two-point boundary value problems involving higher order differential equations", BIT 19, 27-33 (1979).
3.  Elias, U. "Focal points for a linear differential equation whose coefficients are of constant signs", Trans. Amer. Math. Soc. 249, 187-202 (1979).
4.  Fite, W. B. "Concerning the zeros of the solutions of certain differential equations", Trans. Amer. Math. Soc. 19, 341-352 (1918).
5.  Levin, A. Ju. "Some problems bearing on the oscillation of solutions of linear differential equations", Dokl. Akad. Nauk SSSR 148, 512-515 (1963).
6.  Levin, A. Ju. "Distribution of the zeros of solutions of a linear differential equation", Dokl. Akad. Nauk SSSR 156, 1281-1284 (1964).
7.  Muldowney, J. "A necessary and sufficient condition for disfocality", Proc. Amer. Math. Soc. 74, 49-55 (1979).
8.  Nicoletti, O. "Sulle condizione iniziali che determinano gli integrali delle equazioni differenziali ordinarie", Atti. Accad. Sci. Torino A. Sci. Fis. Mat. Natur. 33, 746-759 (1897/98).
9.  Polya, G. "On the mean value theorem corresponding to a given linear homogeneous differential equation", Trans. Amer. Math. Soc. 24, 312-324 (1922).
10. Urabe, M. "The degenerate case of boundary value problems associated with weakly nonlinear differential systems", Pub. Resh. Inst. Math. Sci. Kyoto Univ. 4, 545-584 (1969).
11. Vallée Poussin, C. de la. "Sur l'équation différentielle linéaire du second order", J. Math. Pures Appl. (9)8, 125-144 (1929).
12. Wilder, C. E. "Expansion problems of ordinary linear differential equations with auxiliary conditions at more than two points", Trans. Amer. Math. Soc. 18, 415-422 (1917).

# 3. GREEN'S FUNCTION

Throughout, we shall assume that the condition (2.15) is satisfied. Thus, the existence of the fundamental system of solutions $\{x_i(t)\}$, $1 \leq i \leq n$ of (2.12) satisfying

$$(3.1) \qquad \ell_i[x_j] = \delta_{ij}, \quad 1 \leq i, \ j \leq n$$

is assured. We shall denote by $D_i(t)$, $1 \leq i \leq n$ the cofactor of the element $x_i^{(n-1)}(t)$ in the Wronskian $W(t)$. For convenience we shall write $a = a_0$, $b = a_{n+1}$, $x_0(t) = x_{n+1}(t) = D_0(t) = D_{n+1}(t) = 0$. The square $a \leq t$, $s \leq b$ will be represented by $K$; the same square with straight lines of the form $s = a_i$ rejected from it we shall denote by $K_0$; $K_0$ with rejected diagonal $t = s$ by $K_1$.

Any solution of (2.1) can be written as

$$(3.2) \qquad x(t) = \sum_{i=1}^{n} \alpha_i x_i(t) + \int_{a_0}^{t} \frac{1}{W(s)} \sum_{i=0}^{n+1} D_i(s) x_i(t) f(s) ds.$$

From (2.14) and (3.1), it follows that

$$\alpha_i = A_i - \int_{a_0}^{a_i} \frac{1}{W(s)} D_i(s) f(s) ds, \quad 1 \leq i \leq n.$$

Thus, we find

$$x(t) = \sum_{i=1}^{n} A_i x_i(t) - \sum_{i=0}^{n+1} \int_{a_0}^{a_i} \frac{1}{W(s)} D_i(s) x_i(t) f(s) ds$$

$$+ \int_{a_0}^{t} \frac{1}{W(s)} \sum_{i=0}^{n+1} D_i(s) x_i(t) f(s) ds$$

$$= \sum_{i=1}^{n} A_i x_i(t) - \sum_{k=0}^{n} \int_{a_k}^{a_{k+1}} \frac{1}{W(s)} \sum_{i=k+1}^{n+1} D_i(s) x_i(t) f(s) ds$$

$$+ \int_{a_0}^{t} \frac{1}{W(s)} \sum_{i=0}^{n+1} D_i(s) x_i(t) f(s) ds$$

(3.3)
$$= \sum_{i=1}^{n} A_i x_i(t) + \int_{a_0}^{a_{n+1}} g(t,s) f(s) ds$$

where

(3.4)
$$g(t,s) = \begin{cases} - \dfrac{1}{W(s)} \displaystyle\sum_{i=k+1}^{n+1} D_i(s) x_i(t) & t \leq s \\[4mm] \dfrac{1}{W(s)} \displaystyle\sum_{i=0}^{k} D_i(s) x_i(t) & t \geq s \end{cases}$$

$$a_k < s < a_{k+1}, \quad 0 \leq k \leq n.$$

The function $g(t,s)$ is called the Green's function of the BVP (2.12), (2.16) and is uniquely determined in the square $K$. The following properties of the Green's function $g(t,s)$ are fundamental :

(i)    $\dfrac{\partial^{(i)} g(t,s)}{\partial t^i}$ , $0 \leq i \leq n-2$ are continuous in $K_0$

(ii)   $\dfrac{\partial^{(n-1)} g(t,s)}{\partial t^{n-1}}$ is continuous in $K_1$ and on the diagonal $t = s$

undergoes a discontinuity equal to unity, i.e.,

$$\frac{\partial^{(n-1)} g(t+0,t)}{\partial t^{n-1}} - \frac{\partial^{(n-1)} g(t-0,t)}{\partial t^{n-1}} = 1; \quad t \neq a_i, \quad 1 \leq i \leq n$$

(iii)  $g(t,s)$ as a function of $t$ satisfies (2.12), (2.16) in $K_1$.

Thus, the operator $L$ for the boundary conditions (2.16) has an entirely continuous inverse.

**Lemma 3.1 [1]** The Green's function $g_1(t,s)$ of the BVP

$$x'' = 0$$

(3.5)
$$\alpha_0 x(a_1) - \alpha_1 x'(a_1) = 0$$

$$\beta_0 x(a_2) + \beta_1 x'(a_2) = 0 \quad (a_2 > a_1)$$

exists if $\lambda = \alpha_0 \beta_0 (a_2 - a_1) + \alpha_0 \beta_1 + \beta_0 \alpha_1 \neq 0$, and

(3.6)
$$g_1(t,s) = \frac{1}{\lambda} \begin{cases} (\beta_0 t - \beta_0 a_2 - \beta_1)(\alpha_0 s - \alpha_0 a_1 + \alpha_1) & s \leq t \\ (\beta_0 a_2 - \beta_0 s + \beta_1)(\alpha_0 a_1 - \alpha_0 t - \alpha_1) & t \leq s. \end{cases}$$

Hence, if the constants $\alpha_0$, $\alpha_1$, $\beta_0$ and $\beta_1$ are nonnegative, then $g_1(t,s)$ is nonpositive.

**Lemma 3.2 [9, 14]** The Green's function $g_2(t,s)$ of the BVP

(3.7)
$$x^{(n)} = 0$$

(3.8)
$$x(a_i) = x'(a_i) = \ldots = x^{(k_i)}(a_i) = 0, \; 1 \leq i \leq r \; (\geq 2)$$

$$a_1 < a_2 < \ldots < a_r, \; 0 \leq k_i, \; \sum_{i=1}^{r} k_i + r = n$$

exists and $g_2(t,s)/P(t) > 0$ for $a_1 \leq t \leq a_r$, $a_1 < s < a_r$ where

$$P(t) = \prod_{i=1}^{r} (t - a_i)^{k_i + 1}.$$

**Lemma 3.3 [2, 12]** For the Green's function $g_2(t,s)$ of the BVP (3.7), (3.8) the following equality holds

(3.9)
$$\int_{a_1}^{a_r} |g_2(t,s)| ds = \frac{1}{n!} |P(t)|.$$

Proof. The unique solution of the BVP : $x^{(n)} = 1$, (3.8) is

(3.10)
$$x(t) = \frac{1}{n!} P(t).$$

But, from (3.3) this solution can also be expressed as

(3.11)
$$x(t) = \int_{a_1}^{a_r} g_2(t,s)ds.$$

From (3.10) and (3.11), we have the identity

$$\int_{a_1}^{a_r} g_2(t,s)ds = \frac{1}{n!} P(t).$$

The result (3.9) now follows from Lemma 3.2.

Lemma 3.4 [11]   The Green's function $g_3(t,s)$ of the BVP

(3.12)
$$x^{(2m)} = 0$$

(3.13)
$$x^{(i)}(a_1) = x^{(i)}(a_2) = 0, \ 0 \le i \le m-1$$

is given by

(3.14)   $g_3(t,s) = \dfrac{(-1)^m}{(2m-1)!}$
$$\begin{cases} p^m(t,s) \sum\limits_{j=0}^{m-1} \binom{m-1+j}{j}(t-s)^{m-1-j} q^j(t,s) & s \le t \\[3mm] q^m(t,s) \sum\limits_{j=0}^{m-1} \binom{m-1+j}{j}(s-t)^{m-1-j} p^j(t,s) & t \le s \end{cases}$$

where
$$p(t,s) = \frac{(s-a_1)(a_2-t)}{(a_2 - a_1)} , \quad q(t,s) = \frac{(t-a_1)(a_2-s)}{(a_2 - a_1)} .$$

Lemma 3.5 [3]   The Green's function $g_4(t,s)$ of the BVP : (3.7),

$$x^{(i)}(a_1) = 0, \quad 0 \le i \le k-1 \quad (1 \le k \le n-1 \text{ and fixed})$$

(3.15)

$$x^{(i)}(a_2) = 0, \quad k \le i \le n-1$$

is given by

(3.16) $$g_4(t,s) = \frac{1}{(n-1)!} \begin{cases} \sum_{i=0}^{k-1} \binom{n-1}{i}(t-a_1)^i (a_1-s)^{n-i-1} & s \le t \\ \\ -\sum_{i=k}^{n-1} \binom{n-1}{i}(t-a_1)^i (a_1-s)^{n-i-1} & s \ge t \end{cases}$$

and for $a_1 \le t \le a_2$, $a_1 \le s \le a_2$ the following inequalities hold

(3.17) $$(-1)^{n-k} g_4^{(i)}(t,s) \ge 0, \quad 0 \le i \le k-1$$

(3.18) $$(-1)^{n-i} g_4^{(i)}(t,s) \ge 0, \quad k \le i \le n-1$$

where $g_4^{(i)}(t,s)$ denotes the ith derivative $\dfrac{\partial^i}{\partial t^i} g_4(t,s)$.

**Proof.** The proof of (3.17) is given in [15] and (3.18) follows from the explicit representation

(3.19) $$g_4^{(i)}(t,s) = \frac{1}{(n-i-1)!} \begin{cases} 0 & s \le t \\ \\ -(t-s)^{n-i-1} & s \ge t \end{cases}$$

obtained by differentiating i times (3.16) and the identity

$$(t-s)^{n-1} - \sum_{i=k}^{n-1} \binom{n-1}{i}(t-a_1)^i (a_1-s)^{n-i-1} = \sum_{i=0}^{k-1} \binom{n-1}{i}(t-a_1)^i (a_1-s)^{n-i-1}.$$

**Lemma 3.6 [4]** The Green's function $g_5(t,s)$ of the BVP : (3.7),

$$x^{(i)}(a_1) = 0, \ 0 \le i \le n-2$$

(3.20)

$$x^{(p)}(a_2) = 0 \ (0 \le p \le n-1 \text{ and fixed})$$

is given by

(3.21)

$$g_5(t,s) = \frac{-1}{(n-1)!} \begin{cases} (t-a_1)^{n-1}(\frac{a_2-s}{a_2-a_1})^{n-p-1} - (t-s)^{n-1} & s \le t \\[2ex] (t-a_1)^{n-1}(\frac{a_2-s}{a_2-a_1})^{n-p-1} & s \ge t \end{cases}$$

and for $a_1 \le t \le a_2$, $a_1 \le s \le a_2$ the following inequalities hold

(3.22) $\qquad -g_5^{(i)}(t,s) \ge 0, \quad 0 \le i \le p.$

**Proof.** Inequalities (3.22) hold if and only if

(3.23) $\qquad (t-a_1)^{n-i-1}(\frac{a_2-s}{a_2-a_1})^{n-p-1} \ge (t-s)^{n-i-1}$

for all $a_1 \le s < t \le a_2$. Since $(t-a_1) \ge (t-s)$, $(a_2-a_1) \ge (a_2-s)$ and $(t-a_1)(a_2-s) \ge (t-s)(a_2-a_1)$ it follows that

$$(\frac{t-a_1}{t-s})^{n-i-1} \ge (\frac{a_2-a_1}{a_2-s})^{n-i-1} \ge (\frac{a_2-a_1}{a_2-s})^{n-p-1}$$

which is same as (3.23).

**Lemma 3.7 [4]** The Green's function $g_6(t,s)$ of the BVP : (3.7),

$$x^{(p)}(a_1) = 0 \quad (0 \le p \le n-1 \text{ and fixed})$$

(3.24)

$$x^{(i)}(a_2) = 0 \ , \ 0 \le i \le n-2$$

is given by

$$(3.25) \qquad g_6(t,s) = \frac{(-1)^{n+1}}{(n-1)!} \begin{cases} (a_2-t)^{n-1}(\frac{s-a_1}{a_2-a_1})^{n-p-1} & s \le t \\\\ (a_2-t)^{n-1}(\frac{s-a_1}{a_2-a_1})^{n-p-1}-(s-t)^{n-1} & t \le s \end{cases}$$

and for $a_1 \le t \le a_2$, $a_1 \le s \le a_2$ the following inequalities hold

$$(3.26) \qquad (-1)^{n+i+1} g_6^{(i)}(t,s) \ge 0, \quad 0 \le i \le p.$$

Proof. Inequalities (3.26) hold if and only if

$$(3.27) \qquad (a_2-t)^{n-i-1}(\frac{s-a_1}{a_2-a_1})^{n-p-1} - (s-t)^{n-i-1} \ge 0$$

for all $a_1 \le t < s \le a_2$. Since $(a_2-t) \ge (s-t)$, $(a_2-a_1) \ge (s-a_1)$ and $(a_2-t)(s-a_1) \ge (a_2-a_1)(s-t)$ inequality (3.27) is immediate.

Lemma 3.8 [5]  The Green's function $g_7(t,s)$ of the BVP : (3.12),

$$(3.28) \qquad x^{(2i)}(a_1) = x^{(2i)}(a_2) = 0, \quad 0 \le i \le m-1$$

is given by

$$(3.29) \quad g_7(t,s) = \underbrace{\int_{a_1}^{a_2}\cdots\int_{a_1}^{a_2}}_{(m-1) \text{ times}} g_0(t,s_{m-1})g_0(s_{m-1},s_{m-2})\cdots g_0(s_1,s)ds_1\cdots ds_{m-1}$$

where $g_0(t,s)$ is the Green's function of the BVP : $x'' = 0$, $x(a_1) = x(a_2) = 0$, and for $a_1 \le t \le a_2$, $a_1 \le s \le a_2$ the following inequalities hold

$$(3.30) \qquad (-1)^{m-i} g_7^{(2i)}(t,s) \ge 0, \quad 0 \le i \le m-1.$$

Proof. From Lemma 3.1 the Green's function $g_0(t,s)$ is nonpositive. Thus, the inequalities (3.30) are immediate from the representation

$$g_7^{(2i)}(t,s) = \underbrace{\int_{a_1}^{a_2} \cdots \int_{a_1}^{a_2}}_{(m-i) \text{ times}} g_0(t,s_{m-i-1}) g_0(s_{m-i-1},s_{m-i-2}) \cdots g_0(s_1,s) \times ds_1 \cdots ds_{m-i-1}.$$

## COMMENTS AND BIBLIOGRAPHY

For a very general discussion of Green's function and several interesting results see [6-8, 12, 13, 16]. Following algebraic methods Lemmas 3.2 and 3.3 for r = n have also been proved in [10].

1.  Agarwal, R. P. and Srivastava, U. N. "Generalized two-point boundary value problems", J. Math. Phyl. Sci. 10, 367-373 (1976).
2.  Agarwal, R. P. "An identity for Green's function of multipoint boundary value problems", Proc. Tamil Nadu Acad. Sci. 2, 41-43 (1979).
3.  Agarwal, R. P. and Usmani, R. A. "Iterative methods for solving right focal point boundary value problems", J. Comp. Appl. Math. to appear.
4.  Agarwal, R. P. and Krishnamoorthy, P. R. "Boundary value problems for nth order ordinary differential equations", Bull. Inst. Math. Acad. Sinica 7, 211-230 (1979).
5.  Agarwal, R. P. and Akrivis, G. "Boundary value problems occurring in plate deflection theory", J. Comp. Appl. Math. 8, 145-154 (1982).
6.  Bates, P. W. and Gustafson, G. B. "Green's function inequalities for two-point boundary value problems", Pacific J. Math. 59, 327-343 (1975).
7.  Bates, P. W. and Gustafson, G. B. "Maximization of Green's function over classes of multipoint boundary value problems", SIAM J. Math. Anal. 7, 858-871 (1976).
8.  Beesack, P. R. "On the Green's function of an N-point boundary value problem", Pacific J. Math. 12, 801-812 (1962).
9.  Coppel, W. A. Disconjugacy, Lecture notes in Mathematics 220 Springer-Verlag, New York, 1971.
10. Das, K. M. and Vatsala, A. S. "On Green's function of an n-point boundary value problem", Trans. Amer. Math. Soc. 182, 469-480 (1973).
11. Das, K. M. and Vatsala, A. S. "Green's function for n-n boundary value problem and an analogue of Hartman's result", J. Math. Anal. Appl. 51, 670-677 (1975).

24

12. Gustafson, G. B. "A Green's function convergence principle, with applications to computation and norm estimates", Rocky Mountain J. Math. $\underline{6}$, 457-492 (1976).
13. Levin, A. Ju. "Differential properties of Green's function in a many point boundary value problem", Soviet Math. Dokl. $\underline{2}$, 154-157 (1961).
14. Levin, A. Ju. "Non-oscillation of solutions of the equation $x^{(n)} + p_1(t)x^{(n-1)} + \ldots + p_n(t)x = 0$", Uspehi mat. nauk XXIV, $\underline{2}$(146) 44-96 (1969).
15. Peterson, A. C. "Green's functions for focal type boundary value problems", Rocky Mountain J. Math. $\underline{9}$, 721-732 (1979).
16. Pokornyǐ, Ju. V. "Some estimates of the Green's function of a multi-point boundary value problem", Mat. Zametki $\underline{4}$, 533-540 (1968).

# 4. METHOD OF COMPLEMENTARY FUNCTIONS

Shooting methods, in which the numerical solution of a BVP is found by integrating an appropriate initial value problem, have been the subject of several papers [1-5, 9-11 and references therein] and books [8, 13, 14], and a large part of the monographs [7, 12]. The attraction of these methods is in the availability on most computers of reasonably adequate subroutines of the numerical solution of initial value problems. Goodman and Lance [6] gave two such methods for the numerical solution of linear two point BVPs, the method of complementary functions and the method of adjoints, which are now in common use. Here, we shall formulate the method of complementary functions for the BVP (2.1), (2.2) whereas the method of adjoints is the subject of our next section.

Let $x_j(t)$, $1 \leq j \leq n$ be the solutions of the homogeneous equation (2.12), satisfying

$$(4.1) \qquad x_j^{(i)}(a_1) = \delta_{j-1,i} \; ; \; 1 \leq j \leq n, \; 0 \leq i \leq n-1$$

and $\phi(t)$ be the solution of the nonhomogeneous equation (2.1), satisfying

$$(4.2) \qquad \phi^{(i)}(a_1) = 0, \quad 0 \leq i \leq n-1.$$

Then, any solution of the equation (2.1) can be written as

$$(4.3) \qquad x(t) = \sum_{j=1}^{n} x_j(t) x^{(j-1)}(a_1) + \phi(t).$$

The solution (4.3) is also a solution of the BVP (2.1), (2.2) if and only if

$$(4.4) \qquad \sum_{k=0}^{n-1} c_{ik} [\sum_{j=1}^{n} x_j^{(k)}(a_i) x^{(j-1)}(a_1) + \phi^{(k)}(a_i)] = A_i, \quad 1 \leq i \leq n$$

which can be written as

$$Pu = v$$

where $P$ is an $n \times n$ matrix with elements $p_{ij} = \sum_{k=0}^{n-1} c_{ik} x_j^{(k)}(a_i)$, $u$ is an $n \times 1$ unknown vector with components $u_j = x^{(j-1)}(a_1)$ and $v$ is an $n \times 1$ vector with components $v_i = A_i - \sum_{k=0}^{n-1} c_{ik} \phi^{(k)}(a_i)$.

If the matrix $P$ is nonsingular then the linear system (4.4) can be solved for the unknown vector $u$. Substituting these values of $x^{(j-1)}(a_1)$, $1 \le j \le n$ in (4.3) we find the required solution of (2.1), (2.2).

Thus, to obtain the solution of the BVP (2.1), (2.2) we need to integrate equation (2.12) $n$ times and equation (2.1) only once, i.e., a total of $(n+1)$ integrations are necessary. However, in numerical integration about storing there is some difficulty; besides the values of $x_j^{(k)}(a_i)$, $\phi^{(k)}(a_i)$; $0 \le k \le n-1$, $1 \le j \le n$, $1 \le i \le n$ we need to store the values of $x_j(t)$, $1 \le j \le n$ and $\phi(t)$ at all the grid points, this may not be feasible. A more practical approach is to collect only the values of $x_j^{(k)}(a_i)$ and $\phi^{(k)}(a_i)$ needed in the system (4.4), and solve it for $x^{(j-1)}(a_1)$, $1 \le j \le n$. The solution $x(t)$ is then obtained by integrating (2.1) with these obtained values of $x^{(j-1)}(a_1)$, $1 \le j \le n$. This process of computing the solution of the BVP (2,1), (2.2) is called forward-forward process because all the necessary integrations are performed from the point $a_1$ to $a_n$. Analogous to this method we have backward-backward process in which all the necessary integrations are performed from the point $a_n$ to $a_1$. For this, any solution of (2.1) can also be written as

$$(4.5) \qquad x(t) = \sum_{j=1}^{n} y_j(t) \, x^{(j-1)}(a_n) + \psi(t)$$

where $y_j(t)$ is the solution of the equation (2.12), satisfying

$$(4.6) \qquad y_j^{(i)}(a_n) = \delta_{j-1,i}, \quad 0 \le i \le n-1$$

and $\psi(t)$ is the solution of the equation (2.1), satisfying

$$(4.7) \qquad \psi^{(i)}(a_n) = 0, \quad 0 \le i \le n-1.$$

The solution (4.5) satisfies the boundary conditions (2.2) if and only if

$$(4.8) \qquad \sum_{k=0}^{n-1} c_{ik}[\sum_{j=1}^{n} y_j^{(k)}(a_i) x^{(j-1)}(a_n) + \psi^{(k)}(a_i)] = A_i, \quad 1 \le i \le n.$$

The system (4.8) provides the values of $x^{(j-1)}(a_n)$, $1 \le j \le n$ which we substitute in (4.5) to find the required solution. If only $y_j^{(k)}(a_i)$ and $\psi^{(k)}(a_i)$ needed in the system (4.8) are stored, then we need to integrate (2.1) once more from the obtained values of $x^{(j-1)}(a_n)$, $1 \le j \le n$.

In conclusion the forward-forward process as well as backward-backward process require same number of integrations and theoretically are the same. However, if the problem is numerically instable in forward (backward) direction, then backward-backward (forward-forward) method has to be used.

A modification of the method of complementary functions known as the method of particular solutions is recently proposed in [9-11]. In this method we integrate equation (2.1) with (n+1) different sets of initial conditions

$$z_j^{(i)}(a_1) = \delta_{j-1,i}; \quad 1 \le j \le n, \ 0 \le i \le n-1$$

$$(4.9)$$

$$z_{n+1}^{(i)}(a_1) = 0, \quad 0 \le i \le n-1$$

to obtain $z_j(t)$, $1 \leq j \leq n+1$, i.e., $(n+1)$ particular solutions of (2.1). Next, we introduce $(n+1)$ constants $\alpha_j$, $1 \leq j \leq n+1$ and demand that the linear combination

$$(4.10) \qquad x(t) = \sum_{j=1}^{n+1} \alpha_j \, z_j(t)$$

to be a solution of the BVP (2.1), (2.2). For this, we must have

$$(4.11) \qquad \sum_{j=1}^{n+1} \alpha_j = 1$$

and on substituting (4.10) in (2.2) we get n more equations

$$(4.12) \qquad \sum_{k=0}^{n-1} c_{ik} [ \sum_{j=1}^{n+1} \alpha_j \, z_j^{(k)}(a_i)] = A_i, \ 1 \leq i \leq n.$$

These $(n+1)$ equations (4.11), (4.12) are solved for the $(n+1)$ unknowns $\alpha_j$, $1 \leq j \leq n+1$.

The method of particular solutions is theoretically the same as the method of complementary functions. For this, from (4.11) we have $\alpha_{n+1} = 1 - \sum_{j=1}^{n} \alpha_j$ and hence (4.10) can be written as

$$x(t) = \sum_{j=1}^{n} \alpha_j (z_j(t) - z_{n+1}(t)) + z_{n+1}(t)$$

which is same as (4.3) follows from the fact that $x_j(t) = z_j(t) - z_{n+1}(t)$, $1 \leq j \leq n$ and $\phi(t) = z_{n+1}(t)$. However, it uses only the nonhomogeneous equation (2.1) in contrast with the method of complementary functions where (2.1) as well as the homogeneous equation (2.12) is being used. But, it leads to a system of $(n+1)$ equations instead of n equations.

Next, we shall show that the method of the complementary functions require only $(n-k_1)$ integrations for the BVP (2.1), (2.4) instead of $(n+1)$. For this, we rewrite (4.3) as

$$(4.13) \qquad x(t) = \sum_{j=k_1+2}^{n} x_j(t) x^{(j-1)}(a_1) + \sum_{j=1}^{k_1+1} x_j(t) A_{j,1} + \phi(t)$$

then the expression

$$w_1(t) = \sum_{j=1}^{k_1+1} x_j(t) A_{j,1} + \phi(t)$$

is the solution of the differential equation (2.1), satisfying

$$(4.14) \qquad w_1^{(i)}(a_1) = \begin{cases} A_{i+1,1} & 0 \le i \le k_1 \\ \\ 0 & k_1+1 \le i \le n-1. \end{cases}$$

Hence, (4.13) may be written as

$$(4.15) \qquad x(t) = \sum_{j=k_1+2}^{n} x_j(t) \, x^{(j-1)}(a_1) + w_1(t).$$

Thus, to obtain (4.15) we require $x_j(t)$, $k_1+2 \le j \le n$, i.e., $(n-k_1-1)$ integrations of (2.12) and a particular solution $w_1(t)$ of (2.1) satisfying (4.14).

Substituting (4.15) in (2.4), we obtain

$$(4.16) \qquad \sum_{j=k_1+2}^{n} x_j^{(s)}(a_i) x^{(j-1)}(a_1) + w_1^{(s)}(a_i) = A_{s+1,i}; 0 \le s \le k_i, \ 2 \le i \le r.$$

The $(n-k_1-1)$ equations (4.16) provide $(n-k_1-1)$ unknowns $x^{(j)}(a_1)$, $k_1+1 \le j \le n-1$.

To avoid unnecessary computation we can always assume that $k_1 = \max\limits_{1 \le i \le r} k_i$, otherwise the role of the point $a_1$ with the point $a_j$ where $k_j$ is maximum can be interchanged.

Similarly, for the BVP (2.1), (2.6) we need only $\min\{n-k+1, k+1\}$ integrations instead of (n+1). For this, we can rewrite (4.3) as

$$(4.17) \qquad x(t) = \sum_{j=k+1}^{n} x_j(t) x^{(j-1)}(a_1) + w_2(t)$$

where $w_2(t)$ is the solution of the differential equation (2.1), satisfying

$$(4.18) \qquad w_2^{(i)}(a_1) = \begin{cases} A_i & 0 \le i \le k-1 \\ 0 & k \le i \le n-1. \end{cases}$$

The unknowns $x^{(j)}(a_1)$, $k \le j \le n-1$ are obtained from the system

$$(4.19) \qquad \sum_{j=k+1}^{n} x_j^{(i)}(a_2) x^{(j-1)}(a_1) + w_2^{(i)}(a_2) = A_i, \quad k \le i \le n-1.$$

The solution (4.5) can also be written as

$$(4.20) \qquad x(t) = \sum_{j=1}^{k} y_j(t) x^{(j-1)}(a_2) + w_3(t)$$

where $w_3(t)$ is the solution of the differential equation (2.1), satisfying

$$(4.21) \qquad w_3^{(i)}(a_2) = \begin{cases} 0 & 0 \le i \le k-1 \\ A_i & k \le i \le n-1. \end{cases}$$

The unknowns $x^{(j)}(a_2)$, $0 \le j \le k-1$ are obtained from the system

$$(4.22) \qquad \sum_{j=1}^{k} y_j^{(i)}(a_1) x^{(j-1)}(a_2) + w_3^{(i)}(a_1) = A_i, \quad 0 \le i \le k-1.$$

The BVPs (2.1), (2.7) and (2.1), (2.8) require only 2 integrations, whereas (2.1), (2.9) needs (m+1) integrations. In fact, for n = 2m the solution (4.3) can be written as

$$(4.23) \qquad x(t) = \sum_{j=2(\text{even})}^{2m} x_j(t) x^{(j-1)}(a_1) + w_4(t)$$

where $w_4(t)$ is the solution of the differential equation (2.1), satisfying

$$(4.24) \qquad w_4^{(i)}(a_1) = \begin{cases} A_i & \text{for i even} \\ \\ 0 & \text{for i odd, } 0 \le i \le 2m-1. \end{cases}$$

The unknowns $x^{(i)}(a_1)$; for i odd, $0 \le i \le 2m-1$ are obtained from the system

$$(4.25) \qquad \sum_{j=2(\text{even})}^{2m} x_j^{(2i)}(a_2) x^{(j-1)}(a_1) + w_4^{(2i)}(a_2) = B_{2i}, \quad 0 \le i \le m-1.$$

**Example 4.1**   Consider the three point BVP

$$(4.26) \qquad x''' - x'' + x' - x = t^2 + t$$

$$(4.27) \qquad x(0) = 0, \ x'(\tfrac{\pi}{4}) = 1, \ x''(\tfrac{\pi}{2}) = -2.$$

Since x(0) = 0, from (4.3) any solution of (4.26) can be written as

$$(4.28) \qquad x(t) = x_2(t)x'(0) + x_3(t)x''(0) + \phi(t).$$

An easy  computation provides

$$x_2(t) = \sin t$$

$$x_3(t) = \frac{1}{2} e^t - \frac{1}{2} \cos t - \frac{1}{2} \sin t$$

$$\phi(t) = \frac{3}{2} e^t - \frac{1}{2} \cos t + \frac{3}{2} \sin t - t^2 - 3t - 1$$

thus, the system (4.4) for the BVP (4.26), (4.27) reduces to

$$\frac{1}{\sqrt{2}} x'(0) + \frac{1}{2} \exp(\frac{\pi}{4}) x''(0) + \frac{3}{2} \exp(\frac{\pi}{4}) + \sqrt{2} - \frac{\pi}{2} - 3 = 1$$

(4.29)

$$-x'(0) + \frac{1}{2}[\exp(\frac{\pi}{2}) + 1]x''(0) + \frac{3}{2} \exp(\frac{\pi}{2}) - \frac{3}{2} - 2 = -2.$$

Solving the system (4.29), we obtain the missing $x'(0) = 2.788344412\ldots$ and $x''(0) = -1.007618296\ldots$ .

Example 4.2   Consider the BVP

(4.30)        $x^{(4)} = tx - (11 + 9t + t^2 - t^3)e^t$

(4.31)        $x(-1) = 0$,   $x''(-1) = \frac{2}{e}$,   $x(1) = 0$,   $x''(1) = -6e$

which has a unique solution $x(t) = (1-t^2)e^t$.

From (4.23) any solution of (4.30) can be written as

$$x(t) = x_2(t)x'(-1) + x_4(t)x'''(-1) + w_4(t).$$

Also, the system (4.25) for the BVP (4.30), (4.31) reduces to

$$x_2(1)x'(-1) + x_4(1)x'''(-1) + w_4(1) = 0$$

$$x_2''(1)x'(-1) + x_4''(1)x'''(-1) + w_4''(1) = -6e.$$

The required solutions $x_2(t)$, $x_4(t)$ and $w_4(t)$ are computed by using the Runge-Kutta method of order 4 with step size $h = \frac{b-a}{N}$ . The error

$$\|E\| = \max_{0 \le i \le N} |(1-t_i^2)\exp(t_i) - x(t_i)|, \text{ where } t_i = -1 + ih \text{ and } x(t_i) \text{ is}$$

the computed solution of (4.30), (4.31) is the following

| N  | $\|E\|$  |
|----|----------|
| 8  | 3.1 E-3  |
| 16 | 2.1 E-4  |
| 32 | 1.4 E-5. |

## COMMENTS AND BIBLIOGRAPHY

Method of the complementary functions is one of the easiest method to construct the solution of the linear BVPs. In particular at least for the smooth problems it avoids a lengthy problem preparation required to solve these problems by finite difference methods.

1.   Agarwal, R. P. "The numerical solution of multipoint boundary value problems", J. Comp. Appl. Math. 5, 17-24 (1979).
2.   Agarwal, R. P. "On the periodic solutions of nonlinear second order differential systems", J. Comp. Appl. Math. 5, 117-123 (1979).
3.   Agarwal, R. P. "On the method of complementary functions for non-linear boundary value problems", J. Optimization Theory Appl. 36, 139-144 (1982).
4.   Agarwal, R. P. and Akrivis, G. "Boundary value problems occurring in plate deflection theory", J. Comp. Appl. Math. 8, 145-154 (1982).
5.   Agarwal, R. P. and Usmani, R. A. "Iterative methods for solving right focal point boundary value problems", J. Comp. Appl. Math. to appear.
6.   Goodman, T. R. and Lance, C. N. "The numerical integration of two-point boundary-value problems", Mathematical Tables and Other Aids to Computation 10, 82-86 (1956).
7.   Keller, H. B. Numerical Methods for Two Point Boundary Value Problems, Ginn-Blaisdell, Waltham, 1968.
8.   Meyer, G. H. Initial Value Methods for Boundary Value Problems, Theory and Applications of Invariant Imbedding, Academic Press, New York, 1973.
9.   Miele, A. "Method of particular solutions for linear, two-point boundary-value problems", J. Optimization Theory Appl. 2, 260-273 (1968).

34

10.  Miele, A. and Iyer, R. R."General technique for solving nonlinear, two-point boundary-value problems via the method of particular solution", J. Optimization Theory Appl. 5, 382-399 (1970).
11.  Miele, A. and Iyer, R. R. "Modified quasilinearization method for solving nonlinear, two-point boundary-value problems",J. Math. Anal. Appl. 36, 674-692 (1971).
12.  Na, T. Y. Computational Methods in Engineering Boundary Value Problems, Academic Press, New York, 1979.
13.  Roberts, S. M. and Shipman, J. S. Two-Point Boundary Value Problems: Shooting Methods, Elsevier, New York, 1972.
14.  Scott, M. R. Invariant Imbedding and its Applications to Ordinary Differential Equations : An Introduction, Addision-Wesley, Reading, 1973.

# 5.  METHOD OF ADJOINTS

As the name suggests we use the adjoint equation of (2.1) to obtain the solution of the BVP (2.1), (2.2). However, to avoid unnecessary differentiability conditions on the functions $p_i(t)$, $1 \le i \le n$ appearing in (2.1) we shall formulate this method for the first order differential systems.

Consider the linear nonhomogeneous system of order n

$$(5.1) \qquad\qquad u' = A(t)u + b(t), \ t \ \epsilon \ I$$

where $A(t)$ is an $n \times n$ matrix with elements $a_{ij}(t)$, $1 \le i, j \le n$ which are continuous in I, $b(t)$ is an $n \times 1$ vector with components $b_i(t)$, $1 \le i \le n$ which are continuous in I and $u(t)$ is an unknown vector with components $u_i(t)$, $1 \le i \le n$.

Our concern is about the solution of (5.1), satisfying the boundary conditions

$$(5.2) \qquad\qquad \sum_{k=1}^{N} L_k \ u(a_k) = \ell$$

where $a \le a_1 < a_2 < \ldots < a_N \le b$ $(N \ge 2)$; $L_k$, $1 \le k \le N$ are $n \times n$ given matrices with elements $\alpha_{ij}^k$, $1 \le i, j \le n$ and $\ell$ is an $n \times 1$ vector with components $\ell_i$, $1 \le i \le n$.

The adjoint system for (5.1) is defined by

$$(5.3) \qquad\qquad v' = -A^T(t)v, \ t \ \epsilon \ I$$

where $A^T(t)$ is the transpose of the matrix $A(t)$.

We multiply the ith equation of (5.1) by $v_i(t)$ and sum over all n equations, to obtain

(5.4) $\quad \sum_{i=1}^{n} v_i(t)u_i'(t) = \sum_{i=1}^{n} v_i(t) \sum_{j=1}^{n} a_{ij}(t)u_j(t) + \sum_{i=1}^{n} v_i(t)b_i(t).$

Similarly, we multiply the ith equation of (5.3) by $u_i(t)$ and sum over all n equations, to get

(5.5) $\quad \sum_{i=1}^{n} u_i(t) \, v_i'(t) = - \sum_{i=1}^{n} u_i(t) \sum_{j=1}^{n} a_{ji}(t)v_j(t).$

On adding (5.4) and (5.5), we find

(5.6) $\quad \sum_{i=1}^{n} [v_i(t)u_i'(t) + u_i(t)v_i'(t)] = \sum_{i=1}^{n} v_i(t)b_i(t).$

Next, on integrating (5.6) from $a_1$ to t, we obtain

(5.7) $\quad \sum_{i=1}^{n} v_i(t)u_i(t) - \sum_{i=1}^{n} v_i(a_1)u_i(a_1) = \int_{a_1}^{t} \sum_{i=1}^{n} v_i(s)b_i(s) \, ds.$

Equation (5.7) is called the fundamental identity for the method of adjoints. If V(t) is the fundamental matrix solution of (5.3), satisfying $V(a_N) = E$ (unit $n \times n$ matrix) then, from (5.7) any solution of (5.1) can be written as

(5.8) $\quad u(t) = [V^T(t)]^{-1} V^T(a_1)u(a_1) + [V^T(t)]^{-1} \int_{a_1}^{t} V^T(s)b(s)ds.$

Substituting (5.8) in (5.2), we get

(5.9) $\sum_{k=1}^{N} L_k[V^T(a_k)]^{-1} V^T(a_1)u(a_1) = \ell - \sum_{k=1}^{N} L_k[V^T(a_k)]^{-1} \int_{a_1}^{a_k} V^T(s)b(s)ds$

which is a system of n equations in n unknowns $u_i(a_1)$, $1 \le i \le n$. The system (5.9) has a unique solution provided the matrix $\sum_{k=1}^{N} L_k [V^T(a_k)]^{-1} \times V^T(a_1)$ is nonsingular.

Thus, to find the solution of the BVP (2.1), (2.2) we need to integrate the adjoint system (5.3) n times backward with the terminal conditions

(5.10) $\qquad v_i^j(a_N) = \delta_{ij}$ , $1 \le i$, $j \le n$.

However, the computation of the inverse of $V^T(t)$ at each grid point is a formidable task and we must seek an alternative representation of (5.8). For this, let

$$r(t) = [V^T(t)]^{-1} \int_{a_1}^{t} V^T(s)b(s)ds$$

then, successively it follows that

$$r'(t) = b(t) - [V^T(t)]^{-1} \frac{dV^T(t)}{dt} [V^T(t)]^{-1} \int_{a_1}^{t} V^T(s)b(s)ds$$

$$= b(t) - [V^T(t)]^{-1} \frac{dV^T(t)}{dt} r(t)$$

$$= b(t) - [V^T(t)]^{-1}[- V^T(t)A(t)]r(t)$$

$$= A(t)r(t) + b(t),$$

i.e.,$r(t)$ is a particular solution of (5.1), satisfying the initial condition

(5.11) $\qquad r(a_1) = 0.$

Further, if we denote

$$U(t) = [V^T(t)]^{-1} V^T(a_1)$$

then, $U(a_1) = E$ and $U'(t) = A(t)U(t)$. Hence, $U(t)$ is the fundamental matrix solution of the homogeneous system

(5.12) $\qquad u' = A(t)u.$

Thus, (5.8) can be written as

(5.13) $\qquad u(t) = \sum_{i=1}^{n} u^i(t)u_i(a_1) + r(t)$

where $u^i(t)$ is the solution of (5.12), satisfying

(5.14) $\qquad u_j^i(a_1) = \delta_{ij} , \quad 1 \leq j \leq n$

and $r(t)$ is the solution of (5.1), satisfying (5.11). The unknowns $u_i(a_1)$, $1 \leq i \leq n$ are obtained from the system

(5.15) $\qquad \sum_{k=1}^{N} L_k [ \sum_{i=1}^{n} u^i(a_k)u_i(a_1) + r(a_k)] = \ell.$

However, this is nothing new but the method of the complementary functions for the BVP (5.1), (5.2). Thus, arriving at (5.8) from (5.7) theoretically does not use the adjoint system (5.3) and practically it is almost an impossible task. So, we will use (5.7) directly to obtain the missing initial conditions $u_i(a_1)$, $1 \leq i \leq n$. This will be achieved at the cost of $(N-1)n$ integrations of the adjoint system (5.3).

We integrate (5.3) backward once for each $u_i(a_k)$, $2 \leq k \leq N$ appearing in (5.2) with the terminal conditions

(5.16) $\qquad v_i^{j(k)}(a_k) = \alpha_{ji}^k ; \quad 2 \leq k \leq N, \ 1 \leq i, \ j \leq n$

where $v_i^{j(k)}(a_k)$ is the ith component at $a_k$ for the jth backward solution.

Substituting (5.16) in (5.7), we obtain

$$(5.17) \qquad \sum_{i=1}^{n} \alpha_{ji}^{k} u_i(a_k) - \sum_{i=1}^{n} v_i^{j(k)}(a_1) u_i(a_1)$$

$$= \int_{a_1}^{a_k} \sum_{i=1}^{n} v_i^{j(k)}(s) b_i(s) ds, \quad 2 \le k \le N.$$

Summing (N-1) equations (5.17) and making use of (5.2), we get

$$(5.18) \qquad \sum_{i=1}^{n} [\alpha_{ji}^{1} + \sum_{k=2}^{N} v_i^{j(k)}(a_1)] u_i(a_1)$$

$$= \ell_j - \sum_{k=2}^{N} \int_{a_1}^{a_k} \sum_{i=1}^{n} v_i^{j(k)}(s) b_i(s) ds, \quad 1 \le j \le n.$$

If the matrix $(\alpha_{ji}^{1} + \sum_{k=2}^{N} v_i^{j(k)}(a_1))$ is nonsingular, then the

system (5.18) provides the missing initial conditions $u_i(a_1)$, $1 \le i \le n$. The solution of the BVP (5.1), (5.2) is obtained by integrating forward the system (5.1) with these values of $u_i(a_1)$, $1 \le i \le n$. However, to evaluate the integral term in (5.18) we need to store the solutions of (5.3) not only at the grid points but also at several other intermediate points which have to be interpolated. This besides storage causes inaccuracy and can be avoided at the cost of solving another (N-1) systems. For this, we denote

$$w_{j(k)}(t) = - \int_{t}^{a_k} \sum_{i=1}^{n} v_i^{j(k)}(s) b_i(s) ds; \quad 1 \le j \le n, \; 2 \le k \le N$$

which is equivalent to solving

(5.19)  $$w'_{j(k)}(t) = \sum_{i=1}^{n} v_i^{j(k)}(t) b_i(t)$$

(5.20)  $$w_{j(k)}(a_k) = 0; \quad 1 \le j \le n, \quad 2 \le k \le N.$$

Thus, at the point $a_k$, $2 \le k \le N$ we integrate backward a system of order 2n given by the differential equations (5.3) and (5.19) subject to the conditions (5.16) and (5.20).

With this adjustment system (5.18) takes the form

(5.21)  $$\sum_{i=1}^{n} [\alpha_{ji}^1 + \sum_{k=2}^{N} v_i^{j(k)}(a_1)] u_i(a_1) = \ell_j + \sum_{k=2}^{N} w_{j(k)}(a_1), \quad 1 \le j \le n.$$

This method of constructing the solution of (5.1), (5.2) is called backward-forward process and requires (N-1)n backward integrations of the adjoint system (5.3), satisfying (5.16); (N-1) backward integrations of (5.19), satisfying (5.20) and 1 forward integration of (5.1) with the obtained values of $u_i(a_1)$, $1 \le i \le n$ from the system (5.21), i.e., a total of (N-1)(n+1) + 1 integrations of nth order systems. In particular, if N = 2 then once again we need (n+2) integrations as in forward-forward or backward-backward process.

Similar to backward-forward process we have forward-backward process and for this, we integrate (5.6) from t to $a_N$, to obtain

(5.22)  $$\sum_{i=1}^{n} v_i(a_N) u_i(a_N) - \sum_{i=1}^{n} v_i(t) u_i(t) = \int_t^{a_N} \sum_{i=1}^{n} v_i(s) b_i(s) ds.$$

We integrate (5.3) forward once for each $u_i(a_k)$, $1 \le k \le N-1$ appearing in (5.2) with the initial conditions

(5.23)  $$\underline{v}_i^{j(k)}(a_k) = \alpha_{ji}^k; \quad 1 \le k \le N-1, \quad 1 \le i, j \le n$$

where $v_{-i}^{j(k)}(a_k)$ is the ith component at $a_k$ for the jth forward solution.

Substituting (5.23) in (5.22), we obtain

(5.24)
$$\sum_{i=1}^{n} v_{-i}^{j(k)}(a_N)u_i(a_N) - \sum_{i=1}^{n} \alpha_{ji}^{k} u_i(a_k)$$

$$= \int_{a_k}^{a_N} \sum_{i=1}^{n} v_{-i}^{j(k)}(s)b_i(s)ds, \quad 1 \le k \le N-1.$$

Summing (N-1) equations (5.24) and making use of (5.2), we get

(5.25)
$$\sum_{i=1}^{n} [\alpha_{ji}^{N} + \sum_{k=1}^{N-1} v_{-i}^{j(k)}(a_N)]u_i(a_N)$$

$$= \ell_j + \sum_{k=1}^{N-1} \int_{a_k}^{a_N} \sum_{i=1}^{n} v_{-i}^{j(k)}(s)b_i(s)ds, \quad 1 \le j \le n.$$

We introduce

$$w_{-j(k)}(t) = \int_{a_k}^{t} \sum_{i=1}^{n} v_{-i}^{j(k)}(s)b_i(s)ds$$

which is equivalent to solving

(5.26)
$$w'_{-j(k)}(t) = \sum_{i=1}^{n} v_{-i}^{j(k)}(t)b_i(t)$$

(5.27)
$$w_{-j(k)}(a_k) = 0; \quad 1 \le j \le n, \quad 1 \le k \le N-1.$$

Thus, the system (5.25) is same as

$$(5.28) \quad \sum_{i=1}^{n} [\alpha_{ji}^{N} + \sum_{k=1}^{N-1} v_i^{j}(k)(a_N)] u_i(a_N) = \ell_j + \sum_{k=1}^{N-1} w_{j}(k)(a_N), \ 1 \le j \le n.$$

The solution of the BVP (5.1), (5.2) is obtained by integrating backward the system (5.1) with the obtained values of $u_i(a_N)$, $1 \le i \le n$ from the system (5.28).

Next, we shall consider the system (5.1) together with the boundary conditions

$$(5.29) \quad \sum_{i=1}^{n} c_{j,i-1} u_i(a_j) = A_j, \ 1 \le j \le n.$$

Obviously, the BVP (2.1), (2.2) is included in the problem (5.1), (5.29).

We integrate backward the adjoint system (5.3) n-1 times with the terminal conditions

$$(5.30) \quad v_i^{j}(a_j) = c_{j,i-1} \ ; \ 2 \le j \le n, \ 1 \le i \le n$$

where $v_i^{j}(a_j)$ is the ith component at $a_j$ for the jth backward solution.

Substituting (5.30) in (5.7) and using (5.29), we obtain

$$(5.31) \quad \sum_{i=1}^{n} v_i^{j}(a_1) u_i(a_1) = A_j - \int_{a_1}^{a_j} \sum_{i=1}^{n} v_i^{j}(s) b_i(s) ds, \ 2 \le j \le n.$$

We introduce

$$w_j(t) = - \int_{t}^{a_j} \sum_{i=1}^{n} v_i^{j}(s) b_i(s) ds, \ 2 \le j \le n$$

which is equivalent to solving

$$(5.32) \qquad w_j'(t) = \sum_{i=1}^{n} v_i^j(t) b_i(t)$$

$$(5.33) \qquad w_j(a_j) = 0, \quad 2 \le j \le n.$$

Thus, the system (5.31) can be written as

$$(5.34) \qquad \sum_{i=1}^{n} v_i^j(a_1) u_i(a_1) = A_j + w_j(a_1), \quad 2 \le j \le n.$$

System (5.34) together with the boundary condition at $a_1$, i.e.,

$$(5.35) \qquad \sum_{i=1}^{n} c_{1,i-1} u_i(a_1) = A_1$$

form a system of n equations in n unknowns $u_i(a_1)$, $1 \le i \le n$. The solution of the BVP (5.1), (5.29) is obtained by integrating forward the system (5.1) with these values of $u_i(a_1)$, $1 \le i \le n$.

In practice we couple the adjoint system (5.3) with the equation (5.32) and integrate backward this system of (n+1) equations from the point $a_j$, $2 \le j \le n$ to $a_1$ with the terminal conditions (5.30) and (5.33).

Similarly, in the forward-backward process for the BVP (5.1), (5.29) the unknowns $u_i(a_n)$, $1 \le i \le n$ are computed from the system

$$\sum_{i=1}^{n} v_{-i}^j(a_n) u_i(a_n) = A_j + w_j(a_n), \quad 1 \le j \le n-1$$

$$(5.36)$$

$$\sum_{i=1}^{n} c_{n,i-1} u_i(a_n) = A_n$$

where $v_{-i}^j(t)$ is the ith component at $a_j$ for the jth forward solution of (5.3), satisfying

(5.37)  $\underline{v}_i^j(a_j) = c_{j,i-1}$ ;  $1 \le j \le n-1$, $1 \le i \le n$

and $\underline{w}_j(t)$ is the forward solution of the initial value problem

(5.38)  $$w_j^!(t) = \sum_{i=1}^{n} \underline{v}_i^j(t)b_i(t)$$

(5.39)  $\underline{w}_j(a_j) = 0$,  $1 \le j \le n-1$.

The solution of the BVP (5.1), (5.29) is obtained by integrating backward the system (5.1) with the obtained values of $u_i(a_n)$, $1 \le i \le n$.

Finally, from the above procedure it is clear that like in the method of complementary functions the number of integrations can be reduced if some of the boundary points in (5.29) coincide.

Example 5.1  Consider the BVP

(5.40)  $$\begin{vmatrix} u_1^! \\ u_2^! \\ u_3^! \end{vmatrix} = \begin{vmatrix} 0 & 1 & 0 \\ 0 & 0 & 1 \\ 1 & -1 & 1 \end{vmatrix} \begin{vmatrix} u_1 \\ u_2 \\ u_3 \end{vmatrix} + \begin{vmatrix} 0 \\ 0 \\ t^2+t \end{vmatrix}$$

(5.41)  $u_1(0) = 0$,  $u_2(\frac{\pi}{4}) = 1$,  $u_3(\frac{\pi}{2}) = -2$

which is equivalent to the problem (4.26), (4.27).

For the BVP (5.40), (5.41) the system (5.31) reduces to the following

$$v_2^2(0)u_2(0) + v_3^2(0)u_3(0) = 1 - \int_0^{\pi/4} (s^2+s)v_3^2(s)ds$$

(5.42)

$$v_2^3(0)u_2(0) + v_3^3(0)u_3(0) = -2 - \int_0^{\pi/2} (s^2+s)v_3^3(s)ds.$$

An easy computation provides

$$v^2(t) = \begin{vmatrix} \frac{1}{2} \exp(\frac{\pi}{4} - t) - \frac{1}{\sqrt{2}} \cos t \\[2mm] \frac{1}{\sqrt{2}} (\sin t + \cos t) \\[2mm] \frac{1}{2} \exp(\frac{\pi}{4} - t) - \frac{1}{\sqrt{2}} \sin t \end{vmatrix}, \qquad v^3(t) = \begin{vmatrix} \frac{1}{2} \exp(\frac{\pi}{2} - t) + \frac{1}{2}(\cos t - \sin t) \\[2mm] - \cos t \\[2mm] \frac{1}{2} \exp(\frac{\pi}{2} - t) + \frac{1}{2}(\cos t + \sin t) \end{vmatrix}$$

thus, the system (5.42) is same as (4.29) as it should be.

Example 5.2    We use the backward-forward process to find the solution of the BVP (1.7), (1.8). For this, the necessary systems are integrated by using the Runge-Kutta method of order 4 with step size h = 0.001.

At t = 1, we get

$$u_1(1) = 0.035063 \ (0.035058)$$

$$u_2(1) = -0.015708 \ (-0.015739)$$

$$u_3(1) = 0.000492 \ (0.000503)$$

$$u_4(1) = 0.007136 \ (0.007199)$$

where the values in parentheses are the numerical results given by Meyer [5] using the invariant imbedding method and solving the invariant equations using the Runge-Kutta method of order 4 with the same step size.

Example 5.3    The line method for partial differential equations lies midway between analytical and grid methods. The basis of the method is the substitution of finite differences in place of the derivatives with respect to one independent variable and retention of the deriva- tives with respect to the remaining variables. This approach replaces

a given differential equation by a system of differential equations
with a smaller number of independent variables.

Assume that it is required to integrate the equation

(5.43)    $A(t,s) \dfrac{\partial^2 u}{\partial t^2} + 2B(t,s) \dfrac{\partial^2 u}{\partial t \partial s} + C(t,s) \dfrac{\partial^2 u}{\partial s^2} + D(t,s) \dfrac{\partial u}{\partial t}$

$+ E(t,s)\dfrac{\partial u}{\partial s} + F(t,s)u = G(t,s)$

in the region $\Omega$ (Fig. 5.1). For definiteness we assume that (5.43)
is elliptic. Boundary conditions on the boundary
PQRSP are given.

We draw lines parallel to the t axis,
assuming that the distance between any two
adjacent lines is constant and equal to h.
Assume that the region $\Omega$ be intersected by
the lines $s_k = s_0 + kh$, $0 \le k \le m+1$. We set
$s = s_k$ in the equation and substitute difference ratios for the derivatives with respect to s.

Figure 5.1

For example, we can set

$$\dfrac{\partial u}{\partial s} \bigg|_{s=s_k} = \dfrac{1}{h} [u_{k+1}(t) - u_k(t)]$$

where $u_k(t) = u(t,s_k)$. Similarly,

$$\dfrac{\partial^2 u}{\partial t \partial s} \bigg|_{s=s_k} = \dfrac{1}{h} [u'_{k+1}(t) - u'_k(t)]$$

and

$$\dfrac{\partial^2 u}{\partial s^2} \bigg|_{s=s_k} = \dfrac{1}{h^2} [u_{k+1}(t) - 2u_k(t) + u_{k-1}(t)].$$

We substitute this into the equation (5.43), in which we have already set $s = s_k$ and thus obtain a system of m second order ordinary linear equations in m+2 unknown functions $u_i(t)$, $0 \le i \le m+1$.

In the region $\Omega$, the missing two equations can be obtained from the boundary conditions on the line segments PQ and SR of the boundary. The boundary conditions for the unknown functions $u_i(t)$ can easily be obtained from the boundary conditions for the function $u(t,s)$ on PS and QR.

This second order system can easily be converted to first order linear differential equations with multipoint boundary conditions for which the methods of this section are applicable.

Example 5.4    Consider the Fredholm integral equation

$$(5.44) \qquad x(t) = g(t) + \int_{a_1}^{a_N} K(t,s)x(s)ds$$

and assume that $K(t,s)$ can be approximated by semidegenerate or Green's function type kernel

$$K(t,s) = \begin{cases} \sum_{j=1}^{p(i)} a_{ij}(t)b_{ij}(s) & a_i \le s \le t \le a_{i+1} \\ \\ \sum_{j=1}^{q(i)} c_{ij}(t)d_{ij}(s) & a_i \le t < s \le a_{i+1}, \end{cases}$$

$$1 \le i \le N-1$$

where $p(i)$ and $q(i)$, $1 \le i \le N-1$ are fixed positive integers.

Equation (5.44) can be written as

$$x(t) = g(t) + \sum_{i=1}^{N-1} \left\{ \sum_{j=1}^{p(i)} a_{ij}(t)\phi_{ij}(t) + \sum_{j=1}^{q(i)} c_{ij}(t)\psi_{ij}(t) \right\}$$

where

$$\phi_{ij}(t) = \int_{a_i}^{t} b_{ij}(s)x(s)ds \quad ; \quad 1 \le i \le N-1, \ 1 \le j \le p(i)$$

$$\psi_{ij}(t) = \int_{t}^{a_{i+1}} d_{ij}(s)x(s)ds \ ; \quad 1 \le i \le N-1, \ 1 \le j \le q(i).$$

Thus, to find $x(t)$ we need to solve the multipoint BVP given by

$$\phi'_{ij}(t) = b_{ij}(t)[g(t) + \sum_{i=1}^{N-1} \left\{ \sum_{j=1}^{p(i)} a_{ij}(t)\phi_{ij}(t) + \sum_{j=1}^{q(i)} c_{ij}(t)\psi_{ij}(t) \right\}]$$

$$\psi'_{ij}(t) = -d_{ij}(t)[g(t) + \sum_{i=1}^{N-1} \left\{ \sum_{j=1}^{p(i)} a_{ij}(t)\phi_{ij}(t) + \sum_{j=1}^{q(i)} c_{ij}(t)\psi_{ij}(t) \right\}]$$

(5.45)

$$\phi_{ij}(a_i) = 0; \ 1 \le i \le N-1, \ 1 \le j \le p(i)$$

$$\psi_{ij}(a_{i+1}) = 0; \ 1 \le i \le N-1, \ i \le j \le q(i).$$

The BVP (5.45) can be solved by using the methods of this section.

## COMMENTS AND BIBLIOGRAPHY

The method of adjoints has been used extensively to solve two point BVPs [7,8]. The general formulation for the multipoint BVPs presented here has appeared in [1]. Similar methods for the discrete systems have been proposed recently in [2-4]. The method of lines is of common use in Russian literature [6].

1.  Agarwal, R. P. "The numerical solution of multipoint boundary
    value problems", J. Comp. Appl. Math. $\underline{5}$, 17-24 (1979).
2.  Agarwal, R. P. "On multipoint boundary value problems for discrete
    equations", J. Math. Anal. Appl. $\underline{96}$, 520-534 (1983).
3.  Agarwal, R. P. "Initial-value methods for discrete boundary value
    problems", J. Math. Anal. Appl. $\underline{100}$, 513-529 (1984).
4.  Agarwal, R. P. "Computational methods for discrete boundary value
    problems", Appl. Math. Comp. $\underline{18}$, 15-41 (1986).
5.  Meyer, G. H. Initial Value Methods for Boundary Value Problems,
    Theory and Applications of Invariant Imbedding, Academic Press,
    New York, 1973.
6.  Mikhlin, S. G. and Smolitskiy, K. L. Approximate Methods for
    Solution of Differential and Integral Equations, Elsevier, New
    York, 1967.
7.  Na, T. Y. Computational Methods in Engineering Boundary Value
    Problems, Academic Press, New York, 1979.
8.  Roberts, S. M. and Shipman, J. S. Two-Point Boundary Value Problems:
    Shooting Methods, Elsevier, New York, 1972.

# 6. METHOD OF CHASING

This is another practical shooting method originally developed by Gel'fand and Lokutsiyevskii, however first appeared in English literature only recently [4]. Na [6] has briefly described the method and given different formulations for the different particular cases of (2.1), (2.2). Here, we shall follow [1-3] to provide the general formulation of the method for the BVP (2.1), (2.2). The power of the method is illustrated by solving known Holt's problem.

Since the boundary conditions (2.2) are assumed to be linearly independent at the point $a_i$ at least one of the $c_{ik}$, $0 \leq k \leq n-1$ is not zero. Let $c_{ij} \neq 0$, then at this point $a_i$ the boundary condition (2.2) can be rewritten as

$$(6.1) \qquad x^{(j)}(a_i) = \sum_{\substack{k=0 \\ k \neq j}}^{n-1} d_{ik} \, x^{(k)}(a_i) + \alpha_i \, , \qquad 1 \leq i \leq n$$

where

$$d_{ik} = -\frac{c_{ik}}{c_{ij}} \, ; \quad 0 \leq k \leq n-1, \quad k \neq j$$

and

$$\alpha_i = \frac{A_i}{c_{ij}} \, .$$

In the differential equation (2.1), we begin with the assumption that $p_1(t) \equiv 0$, so that

$$(6.2) \qquad x^{(n)} = -\sum_{i=2}^{n} p_i(t) x^{(n-i)} + f(t).$$

Now, for the boundary condition (6.1) we assume that the solution $x(t)$ of (6.2) satisfies $(n-1)$th order linear differential equation

$$(6.3) \qquad x^{(j)}(t) = \sum_{\substack{k=0 \\ k \neq j}}^{n-1} d_{ik}(t)x^{(k)}(t) + \alpha_i(t)$$

where the n functions $d_{ik}(t)$; $0 \leq k \leq n-1$, $k \neq j$ and $\alpha_i(t)$ are to be determined.

Differentiating (6.3) once, we get

$$(6.4) \qquad x^{(j+1)}(t) = \sum_{\substack{k=0 \\ k \neq j}}^{n-1} [d_{ik}(t)x^{(k+1)}(t) + d'_{ik}(t)x^{(k)}(t)] + \alpha'_i(t).$$

Next, we shall use (6.3) to eliminate the term $x^{(n-1)}(t)$ from (6.4), however it depends on a particular value of j and we need to consider four different cases :

(i)   $j = 0$, $n \geq 3$

From (6.3), we have

$$(6.5) \qquad x^{(n-1)}(t) = \frac{1}{d_{i,n-1}(t)}[x(t) - \sum_{k=1}^{n-2} d_{ik}(t)x^{(k)}(t) - \alpha_i(t)].$$

Using (6.5) in (6.4) and rearranging the terms, we get

$$x^{(n)}(t) = - \frac{[d_{i,n-2}(t)+d'_{i,n-1}(t)]}{d^2_{i,n-1}(t)} x(t)$$

$$+ [\frac{1}{d_{i,n-1}(t)} + \frac{d_{i,n-2}(t)+d'_{i,n-1}(t)}{d^2_{i,n-1}(t)} d_{i1}(t) - \frac{d'_{i1}(t)}{d_{i,n-1}(t)}]x'(t)$$

$$+ \sum_{k=2}^{n-2} [\frac{d_{i,n-2}(t)+d'_{i,n-1}(t)}{d^2_{i,n-1}(t)} d_{ik}(t) - \frac{d_{i,k-1}(t)+d'_{ik}(t)}{d_{i,n-1}(t)}]x^{(k)}(t)$$

$$(6.6) \qquad + [\frac{d_{i,n-2}(t)+d'_{i,n-1}(t)}{d^2_{i,n-1}(t)} \; \alpha_i(t) \; - \; \frac{\alpha'_i(t)}{d_{i,n-1}(t)}].$$

Comparing (6.2) and (6.6), we find the system of n differential equations

$$d'_{i,n-1}(t) = -d_{i,n-2}(t) + p_n(t)d^2_{i,n-1}(t)$$

$$d'_{ik}(t) = p_{n-k}(t)d_{i,n-1}(t)-d_{i,k-1}(t)+p_n(t)d_{i,n-1}(t)d_{ik}(t);$$

$$k = n-2,n-3,\ldots,2$$

$(6.7) \quad d'_{i1}(t) = 1 + p_n(t)d_{i,n-1}(t)d_{i1}(t) + p_{n-1}(t)d_{i,n-1}(t)$

$$\alpha'_i(t) = -f(t)d_{i,n-1}(t) + p_n(t)d_{i,n-1}(t)\alpha_i(t).$$

We also desire that this solution x(t) must satisfy the boundary condition (6.1). For this, we compare (6.1) and (6.3) at the point $a_i$ and find

$$d_{ik}(a_i) = d_{ik} \; , \quad 1 \le k \le n-1$$

(6.8)

$$\alpha_i(a_i) = \alpha_i \; .$$

In the rest we proceed as for the case j = 0 and obtain the following systems

(ii)   $1 \le j \le n-3$

$$d'_{i,n-1}(t) = -d_{i,n-2}(t)-d_{i,j-1}(t)d_{i,n-1}(t)+p_{n-j}(t)d^2_{i,n-1}(t)$$

$$d'_{ik}(t) = -d_{i,k-1}(t)-d_{i,j-1}(t)d_{ik}(t)+(p_{n-k}(t)+p_{n-j}(t)d_{ik}(t))d_{i,n-1}(t);$$

$$k = n-2,n-3,\ldots,1, \; k \neq j, \; j+1$$

(6.9)

$$d'_{i,j+1}(t) = 1-d_{i,j-1}(t)d_{i,j+1}(t)+(p_{n-j-1}(t)+p_{n-j}(t)d_{i,j+1}(t))d_{i,n-1}(t)$$

$$d'_{i0}(t) = -d_{i,j-1}(t)d_{i0}(t)+(p_n(t)+p_{n-j}(t)d_{i0}(t))d_{i,n-1}(t)$$

$$\alpha'_i(t) = -d_{i,j-1}(t)\alpha_i(t)+(p_{n-j}(t)\alpha_i(t)-f(t))d_{i,n-1}(t)$$

$$d_{ik}(a_i) = d_{ik}; \quad 0 \le k \le n-1, \; k \ne j$$

(6.10)

$$\alpha_i(a_i) = \alpha_i.$$

(iii)  $j = n-2$

$$d'_{i,n-1}(t) = 1 - d_{i,n-1}(t)d_{i,n-3}(t) + p_2(t)d^2_{i,n-1}(t)$$

$$d'_{ik}(t) = -d_{i,k-1}(t)+(p_{n-k}(t)+p_2(t)d_{ik}(t))d_{i,n-1}(t)-d_{i,n-3}(t)d_{ik}(t),$$

$$1 \le k \le n-3$$

(6.11)

$$d'_{i0}(t) = -d_{i,n-3}(t)d_{i,0}(t)+(p_n(t)+p_2(t)d_{i,0}(t))d_{i,n-1}(t)$$

$$\alpha'_i(t) = -d_{i,n-3}(t)\alpha_i(t)+(-f(t)+p_2(t)\alpha_i(t))d_{i,n-1}(t)$$

$$d_{ik}(a_i) = d_{ik}; \quad 0 \le k \le n-1, \; k \ne n-2$$

(6.12)

$$\alpha_i(a_i) = \alpha_i.$$

(iv)  $j = n-1$

$$d'_{ik}(t) = -d_{i,k-1}(t) - d_{i,n-2}(t)d_{ik}(t) - p_{n-k}(t), \; 1 \le k \le n-2$$

$$d'_{i0}(t) = -d_{i,n-2}(t)d_{i0}(t) - p_n(t)$$

(6.13)

$$\alpha'_i(t) = -d_{i,n-2}(t)\alpha_i(t) + f(t)$$

$$d_{ik}(a_i) = d_{ik}, \; 0 \le k \le n-2$$

(6.14)

$$\alpha_i(a_i) = \alpha_i.$$

For the particular value of j, we integrate the above appropriate system from the point $a_i$ to $a_n$ and collect the values of $d_{ik}(a_n)$; $0 \le k \le n-1$, $k \ne j$ and $\alpha_i(a_n)$. Thus, (6.3) provides a new boundary relation at the point $a_n$

$$(6.15) \qquad x^{(j)}(a_n) = \sum_{\substack{k=0 \\ k \ne j}}^{n-1} d_{ik}(a_n)x^{(k)}(a_n) + \alpha_i(a_n).$$

Let N be the number of different boundary points, i.e., $a_1 < a_2 < \ldots < a_N = a_n$ ($n \ge N \ge 2$) and $m(a_j)$ represents the number of boundary relations (6.1) prescribed at the point $a_j$ and hence $\sum_{j=1}^{N} m(a_j) = n$. Thus, in (6.1) we have $m(a_n)$ boundary relations at the point $a_n$ and to find $x^{(j)}(a_n)$, $0 \le j \le n-1$ we need $n-m(a_n)$ more new relations (6.15), i.e., we need to integrate $n-m(a_n)$ appropriate differential systems.

Finally, from the obtained values of $x^{(j)}(a_n)$, $0 \le j \le n-1$ we integrate backward differential equation (6.2) and obtain the required solution.

With the help of following guidelines unnecessary computation can be avoided : (1) $m(a_n) = \max_{1 \le j \le N} m(a_j)$, otherwise the role of the point $a_n$ with the point $a_j$ where $m(a_j)$ is maximum can be interchanged. (2) We need to integrate $n-m(a_n)$ times but not necessarily different differential systems, specially because differential system does not change as long as in (6.1) j is same. In fact, we can have at most n different differential systems.

For the case $p_1(t) \ne 0$, we rewrite the differential equation (2.1) as

$$(6.16) \qquad [P(t)x^{(n-1)}]' = -\sum_{i=2}^{n} P(t)p_i(t)x^{(n-i)} + P(t)f(t)$$

where $P(t) = \exp(\int_{a_1}^{t} p_1(s)ds).$

Assumption that the solution of (6.16) should satisfy (n-1)th order linear differential equation

$$(6.17) \qquad d_{ij}(t)x^{(j)}(t) = \sum_{\substack{k=0 \\ k \neq j}}^{n-1} d_{ik}(t)x^{(k)}(t) + \alpha_i(t)$$

with $d_{i,n-1}(t) = P(t)$ brings the problem in the realm of the foregoing analysis.

Example 6.1   The two point BVP

$$(6.18) \qquad x'' = (2m + 1 + t^2)x$$

$$(6.19) \qquad x(0) = \beta, \ x(\infty) = 0$$

where m and $\beta$ are specified constants, now known as Holt's problem [5] is a typical example where usual shooting methods fail [9]. Faced with this difficulty Holt [5] used a finite difference method, whereas Osborne [7] used multiple shooting method and Roberts and Shipman [8,9] used a multipoint approach.

For this problem the solution representation (6.3) reduces to

$$(6.20) \qquad x(t) = d_{01}(t)x'(t) + \alpha_0(t)$$

and the case (iii) provides the differential system to be integrated

$$d_{01}'(t) = 1 - (2m+1+t^2)d_{01}^2(t)$$

$$(6.21)$$

$$\alpha_0'(t) = - (2m+1+t^2)d_{01}(t)\alpha_0(t)$$

together with the initial conditions

(6.22) $\qquad d_{01}(0) = 0, \quad \alpha_0(0) = \beta.$

We use fourth order Runge-Kutta method with step size 0.01 and obtain $d_{01}(t)$, $\alpha_0(t)$ at $t = 18.01$. These values are used to calculate $x'(18.01)$ from (6.20). The differential equation (6.18) is integrated backward with the given $x(18.01) = 0$ and the obtained value of $x'(18.01)$ using fourth order Runge-Kutta method with the same step size. The value $t = 18.01$ has been chosen in view of restricted computer capabilities.

The solution thus obtained has been presented in Tables 1-3 for different choices of m and $\beta$. These tables also contain solution of the problem obtained earlier in [5, 7-9].

Table 1.　$m = 0$, $\beta = 1$

| t | Present Solution | Complementary Functions [9] | Finite Differences [5] | Solution by Osborne [7] | Roberts and Shipman [8] |
|---|---|---|---|---|---|
| 0 | 0.9999876 E 00 | 0.10000000 E 01 | 0.100000 E 01 | 0.1000 E 01 | 0.10000000 E 01 |
| 1 | 0.2593404 E 00 | 0.15729920 E 00 | 0.157300 E 00 | 0.2593 E 00 | 0.15729921 E 00 |
| 2 | 0.3456397 E-01 | 0.46777349 E-02 | 0.467778 E-02 | 0.3455 E-01 | 0.46777350 E-02 |
| 3 | 0.1988532 E-02 | 0.22090497 E-04 | 0.220908 E-04 | 0.1987 E-02 | 0.22090497 E-04 |
| 4 | 0.4595871 E-04 | 0.15417257 E-07 | 0.154175 E-07 | 0.4590 E-04 | 0.15417259 E-07 |
| 5 | 0.4125652 E-06 | 0.15366706 E-11 | 0.153749 E-11 | 0.4188 E-06 | 0.15374602 E-11 |
| 6 | 0.1413020 E-08 | -0.73163560 E-15 | 0.215201 E-16 | 0.1409 E-08 | 0.21519753 E-16 |
| 7 | 0.1827268 E-11 | -0.75311525 E-15 | 0.418390 E-22 | 0.1821 E-11 | 0.41838334 E-22 |
| 8 | 0.8863389 E-15 | -0.75515520 E-15 | 0.112244 E-28 | 0.8825 E-15 | 0.11224343 E-28 |
| 9 | 0.1605597 E-18 | | 0.413703 E-36 | 0.1597 E-18 | 0.41370659 E-36 |
| 10 | 0.1082885 E-22 | | 0.208844 E-44 | 0.1058 E-22 | 0.20895932 E-44 |
| 11 | 0.2713141 E-27 | | 0.144078 E-53 | | 0.12279100 E-49 |
| 12 | 0.2521085 E-32 | | 0.135609 E-63 | | 0.13487374 E-49 |
| 13 | 0.8677126 E-38 | | | | 0.17299316 E-60 |
| 14 | 0.1105113 E-43 | | | | -0.25496486 E-65 |
| 15 | 0.5203999 E-50 | | | | |
| 16 | 0.9055032 E-57 | | | | |
| 17 | 0.5818867 E-64 | | | | |
| 18 | 0.4179442 E-72 | | | | |

58

Table 2.  $m = 1$, $\beta = \pi^{-\frac{1}{2}}$

| t | Present Solution | Complementary Functions [9] | Finite Differences [5] |
|---|---|---|---|
| 0 | 0.5641878 E 00 | 0.56418960 E 00 | 0.5642 E 00 |
| 1 | 0.8285570 E-01 | 0.50254543 E-01 | 0.5026 E-01 |
| 2 | 0.7226698 E-02 | 0.97802274 E-03 | 0.9782 E-03 |
| 3 | 0.3020138 E-03 | 0.33550350 E-05 | 0.3356 E-05 |
| 4 | 0.5431819 E-05 | 0.18221222 E-08 | 0.1823 E-08 |
| 5 | 0.3975088 E-07 | 0.12567523 E-12 | 0.1482 E-12 |
| 6 | 0.1146879 E-09 | -0.29349128 E-13 | 0.1747 E-17 |
| 7 | 0.1279827 E-12 | -0.34242684 E-13 | 0.2931 E-23 |
| 8 | 0.5456289 E-16 | -0.39134491 E-13 | 0.6912 E-30 |
| 9 | 0.8813160 E-20 | | |
| 10 | 0.5361614 E-24 | | |
| 11 | 0.1223266 E-28 | | |
| 12 | 0.1043287 E-33 | | |
| 13 | 0.3317918 E-39 | | |
| 14 | 0.3926980 E-45 | | |
| 15 | 0.1727057 E-51 | | |
| 16 | 0.2818780 E-58 | | |
| 17 | 0.1705581 E-65 | | |
| 18 | 0.1160366 E-73 | | |

Table 3.  $m = 2$, $\beta = \frac{1}{4}$

| t | Present Solution | Complementary Functions [9] | Finite Differences [5] |
|---|---|---|---|
| 0 | 0.2500006 E 00 | 0.25000000 E 00 | 0.2500 E 00 |
| 1 | 0.2340787 E-01 | 0.14197530 E-01 | 0.1420 E-01 |
| 2 | 0.1414359 E-02 | 0.19141103 E-03 | 0.1914 E-03 |
| 3 | 0.4411547 E-04 | 0.49007176 E-06 | 0.4901 E-06 |
| 4 | 0.6261059 E-06 | 0.20999802 E-08 | 0.2101 E-08 |
| 5 | 0.3764660 E-08 | -0.36865462 E-13 | 0.1403 E-13 |
| 6 | 0.9193294 E-11 | -0.72849101 E-13 | 0.1400 E-18 |
| 7 | 0.8879995 E-14 | -0.98795539 E-13 | 0.2034 E-24 |
| 8 | 0.3334327 E-17 | -0.12873356 E-12 | 0.4224 E-31 |
| 9 | 0.4809239 E-21 | | |
| 10 | 0.2641945 E-25 | | |
| 11 | 0.5493305 E-30 | | |
| 12 | 0.4302831 E-35 | | |
| 13 | 0.1265030 E-40 | | |
| 14 | 0.1391954 E-46 | | |
| 15 | 0.5719099 E-53 | | |
| 16 | 0.8757835 E-60 | | |
| 17 | 0.4990734 E-67 | | |
| 18 | 0.3216635 E-75 | | |

## COMMENTS AND BIBLIOGRAPHY

Method of chasing offers a very interesting alternative for the conversion of a given linear BVP to its equivalent initial value problem. Under certain conditions on the functions $p_i(t)$, $1 \le i \le n$ the solutions of the associated systems increase slowly with t even though the solution of the original equation (2.1) does increase rapidly. Thus, it is possible to use larger intervals of the independent variable for the same occuracy.

1.  Agarwal, R. P. "On Gel'fand's method of chasing for solving multipoint boundary value problems",Proc. Equadiff 6, Ed. J. Vosmanský, Lecture notes in Mathematics, Springer-Verlag, Berlin (to appear).
2.  Agarwal, R. P. and Gupta, R. C. "On the solution of Holt's problem", BIT 24, 342-346 (1984).
3.  Agarwal, R. P. and Gupta, R. C. "Method of chasing for multipoint boundary value problems", Appl. Math. Comp. 17, 37-43 (1985).
4.  Berezin, I. S. and Zhidkov, N. P. Method of Chasing in Computing Methods (O. M. Blum and A. D. Booth, trans.), Vol. II, Pergamon, Oxford, 1965.
5.  Holt, J. F. "Numerical solution of nonlinear two-point boundary value problems by finite-difference methods", Comm. Asso. Comp. Machinery 7, 366-373 (1964).
6.  Na, T. Y. Computational Methods in Engineering Boundary Value Problems, Academic Press, New York, 1979.
7.  Osborne, M. R. "On shooting methods for boundary value problems", J. Math. Anal. Appl. 27, 417-433 (1969).
8.  Roberts, S. M. and Shipman, J. S. "Multipoint solution of two-point boundary-value problems", J. Optimization Theory Appl. 7, 301-318 (1971).
9.  Roberts, S. M. and Shipman, J. S. Two-Point Boundary Value Problems : Shooting Methods, Elsevier, New York, 1972.

## 7. SECOND ORDER EQUATIONS

The BVP

(7.1) $$x'' = f(t, x, x')$$

(7.2) $$x(a_1) = A, \quad x(a_2) = B$$

where $f(t,x_0,x_1)$ is continuous and satisfies a uniform Lipschitz condition

(7.3) $$|f(t,x_0,x_1) - f(t,y_0,y_1)| \leq L_0|x_0-y_0| + L_1|x_1-y_1|$$

on $[a_1,a_2] \times R^2$ has a long history going back to Picard [11] 1893 (from the literature it appears that before 1893 the main attack was to construct the solution to a given problem tacitly assuming existence and uniqueness). He proved that if $(a_2-a_1)$ is sufficiently small then the sequence $\{x_m(t)\}$ of functions generated by the iterative scheme

$$x''_{m+1}(t) = f(t,x_m(t),x'_m(t))$$

$$x_{m+1}(a_1) = A, \quad x_{m+1}(a_2) = B ; \quad m = 0,1,\ldots$$

with $x_0(t)$ known, converges to the solution of the BVP (7.1), (7.2). In this way he obtained existence and uniqueness over all intervals $[a_1,a_2]$ of length less than h where

$$\tfrac{1}{2} L_0 h^2 + L_1 h < 1.$$

By sharpening the estimates employed in his iterative procedure he later [12] obtained the inequality

(7.4) $$\tfrac{1}{8} L_0 h^2 + \tfrac{1}{2} L_1 h < 1$$

which was further improved by Lettenmeyer [9] to

$$\frac{1}{\pi^2} L_0 h^2 + \frac{4}{\pi^2} L_1 h < 1.$$

In the special case when $f(t,x,x')$ is linear in $x$ and $x'$ and $f(t,0,0)=0$, Opial [10] obtained the condition

$$\frac{1}{\pi^2} L_0 h^2 + \frac{2}{\pi^2} L_1 h < 1$$

and showed that it is best possible in the sense that the coefficients $\frac{1}{\pi^2}$ and $\frac{2}{\pi^2}$ cannot be replaced by smaller ones. But the claim is not true if $L_1 \neq 0$ [2].

For the equation (7.1), Bailey et. al. [3] have also treated the boundary conditions

(7.5)                    $x(a_1) = A, \quad x'(a_2) = B$

and

(7.6)                    $x'(a_1) = A, \quad x(a_2) = B$

and obtained the best possible result for each of the BVPs (7.1), (7.2); (7.1), (7.5); (7.1), (7.6) which are contained in the following :

Theorem 7.1   Let $f \in C[[a_1,a_2] \times R^2, R]$ and satisfy the Lipschitz condition (7.3). Let $y(t)$ be any solution of $y'' + L_1 y' + L_0 y = 0$ which vanishes at $t = a_1$ and let $\alpha(L_1,L_0)$ be the first positive number such that $y'(t) = 0$ at $t = a_1 + \alpha(L_1,L_0)$. Then,

(i) if $a_2 - a_1 < \alpha(L_1,L_0)$ each of the BVPs (7.1), (7.5) and (7.1), (7.6) has a unique solution,

(ii) if $a_2 - a_1 < 2\alpha(L_1,L_0)$ the BVP (7.1), (7.2) has a unique solution.

Thus the best possible existence and uniqueness interval for the BVP (7.1), (7.2) is obtained from the best possible existence and uniqueness interval for the BVPs (7.1), (7.5) and (7.1), (7.6).

A more general boundary conditions which include (7.2), (7.5) and (7.6) are

(7.7)
$$\alpha_0 x(a_1) - \alpha_1 x'(a_1) = A$$

(7.8)
$$\beta_0 x(a_2) + \beta_1 x'(a_2) = B$$

where $\lambda = \alpha_0 \beta_0 (a_2 - a_1) + \alpha_0 \beta_1 + \beta_0 \alpha_1 \neq 0$.

For the BVP (7.1), (7.7), (7.8) first we shall provide merely existence results and then only uniqueness result and finally existence and uniqueness results.

Theorem 7.2 [1]    Let $f \in C[[a_1, a_2] \times R^2, R]$ and $M > 0$, $N > 0$ be given real numbers and Q be the maximum of $|f(t, x_0, x_1)|$ on the compact set

$$\{(t, x_0, x_1) : a_1 \leq t \leq a_2, |x_0| \leq 2M, |x_1| \leq 2N\}.$$

Further, we assume that

$$QK_1 \leq M, \quad QK_2 \leq N$$

$$\frac{1}{|\lambda|} |A\beta_0 (a_2 - a_1) + A\beta_1 + B\alpha_1| \leq M$$

(7.9)
$$\frac{1}{|\lambda|} |B\alpha_0 (a_2 - a_1) + A\beta_1 + B\alpha_1| \leq M$$

$$\frac{1}{|\lambda|} |B\alpha_0 - A\beta_0| \leq N$$

where

$$K_1 = \frac{(a_2-a_1)}{2|\lambda|}[\{(|\alpha_0\beta_0|(a_2-a_1) + 2(|\alpha_1\beta_0|+|\alpha_0\beta_1|))r_1r_2 + |\alpha_0\beta_1|r_1^2$$

$$+ |\alpha_1\beta_0|r_2^2\}(a_2-a_1) + 2|\alpha_1\beta_1|]$$

$$K_2 = \frac{1}{2|\lambda|}[|\alpha_0\beta_0|(a_2-a_1) + 2\max\{|\alpha_1\beta_0|, |\beta_1\alpha_0|\}](a_2-a_1)$$

$$r_1 = \frac{1}{2\lambda^*}[|\alpha_0\beta_0|(a_2-a_1) + 2|\alpha_0\beta_1|]$$

$$r_2 = \frac{1}{2\lambda^*}[|\alpha_0\beta_0|(a_2-a_1) + 2|\alpha_1\beta_0|]$$

and

$$\lambda^* = |\alpha_0\beta_0|(a_2-a_1) + |\alpha_0\beta_1| + |\alpha_1\beta_0|.$$

Then, the BVP (7.1), (7.7), (7.8) has a solution.

Proof. The set

$$B[a_1,a_2] = \{x(t) \in C^{(1)}[a_1,a_2] : \|x\| \leq 2M, \|x'\| \leq 2N\}$$

where $\|\cdot\| = \max_{a_1 \leq t \leq a_2} |\cdot(t)|$ is a closed convex subset of the Banach space $C^{(1)}[a_1,a_2]$. The mapping $T : C^{(1)}[a_1,a_2] \to C^{(2)}[a_1,a_2]$ defined by

$$(7.10) \qquad (Tx)(t) = \ell(t) + \int_{a_1}^{a_2} g_1(t,s)f(s,x(s),x'(s))ds$$

is completely continuous, where $g_1(t,s)$ is defined in (3.6) and $\ell(t)$ is the straight line

$$\ell(t) = \frac{1}{\lambda}[A\beta_0(a_2-t) + A\beta_1 + B\alpha_1 + B\alpha_0(t-a_1)].$$

For $x(t) \in B[a_1,a_2]$ an easy computation provides

$$|(Tx)(t)| \leq \max_{a_1 \leq t \leq a_2} |\ell(t)| + K_1 Q$$

and

$$|(Tx)'(t)| \leq \max_{a_1 \leq t \leq a_2} |\ell'(t)| + K_2 Q.$$

Thus condition (7.9) implies that T maps $B[a_1,a_2]$ into itself. It then follows from the Schauder fixed point theorem that T has a fixed point in $B[a_1,a_2]$. This fixed point is a solution of the stated BVP.

**Corollary 7.3**   Assume that $f \in C[[a_1,a_2] \times R^2, R]$ and

$$|f(t,x_0,x_1)| \leq c_0 + c_1 |x_0|^\alpha + c_2 |x_1|^\beta$$

where $c_0$, $c_1$, $c_2$, $\alpha$, $\beta$ are nonnegative constants and $\alpha < 1$, $\beta < 1$. Then, the BVP (7.1), (7.7), (7.8) has a solution.

Along with the assumption $\lambda \neq 0$, hereafter we shall assume that the constants $\alpha_0$, $\alpha_1$, $\beta_0$ and $\beta_1$ to be nonnegative.

**Definition 7.1**   We call a function $p \in C^{(2)}[a_1,a_2]$ a <u>lower solution</u> of (7.1), (7.7), (7.8) provided that

(7.11)     $p''(t) \geq f(t,p(t),p'(t)), \ t \in [a_1,a_2]$

(7.12)     $\alpha_0 p(a_1) - \alpha_1 p'(a_1) \leq A$

(7.13)     $\beta_0 p(a_2) + \beta_1 p'(a_2) \leq B.$

Similarly, a function $q \in C^{(2)}[a_1,a_2]$ is called an <u>upper solution</u> of (7.1), (7.7), (7.8) if

(7.14)     $q''(t) \leq f(t,q(t),q'(t)), t \in [a_1,a_2]$

(7.15) $$\alpha_0 q(a_1) - \alpha_1 q'(a_1) \geq A$$

(7.16) $$\beta_0 q(a_2) + \beta_1 q'(a_2) \geq B.$$

Theorem 7.4 [1]    Let the conditions of Corollary 7.3 be satisfied. Further, we assume that there exist p, q lower and upper solutions of (7.1), (7.7), (7.8) satisfying $p(t) \leq q(t)$ on $[a_1,a_2]$. Then, there exists a solution x(t) of the BVP (7.1), (7.7), (7.8) such that $p(t) \leq x(t) \leq q(t)$ on $[a_1,a_2]$.

Proof.  Define the function F(t,x,x') on $[a_1,a_2] \times R^2$ by setting

$$F(t,x,x') = f(t,q(t),x') + \frac{x - q(t)}{1 + x^2} \qquad \text{if } x > q(t)$$

$$= f(t,x,x') \qquad \text{if } p(t) \leq x \leq q(t)$$

$$= f(t,p(t),x') - \frac{p(t) - x}{1 + x^2} \qquad \text{if } x < p(t).$$

Since f satisfies the conditions of Corollary 7.3, F also does. Hence, there exists a solution x(t) of the differential equation $x'' = F(t,x,x')$ which satisfies the boundary conditions (7.7), (7.8).

We shall now show that $p(t) \leq x(t) \leq q(t)$ on $[a_1,a_2]$, this means that x(t) is a solution of (7.1), (7.7), (7.8). Assume that $p(t) > x(t)$ at some point of $[a_1,a_2]$, then $p(t) - x(t)$ attains a positive maximum at some point $t_0 \in [a_1,a_2]$. Suppose $t_0 = a_1$, then $p'(a_1) - x'(a_1) \leq 0$ and hence if $\alpha_0 \neq 0$ the condition (7.12) is violated, also if $\alpha_0 = 0$ then (7.12) is satisfied only if $p'(a_1) - x'(a_1) = 0$. Similarly, if $t_0 = a_2$ then condition (7.13) holds only if $p'(a_2) - x'(a_2) = 0$. Now at the positive maximum point $t_0$, we have

$$p''(t_0) - x''(t_0) \geq f(t_0,p(t_0), p'(t_0)) - F(t_0,x(t_0),x'(t_0))$$

$$= f(t_0, p(t_0), p'(t_0)) - f(t_0, p(t_0), x'(t_0)) + \frac{p(t_0) - x(t_0)}{1 + x^2(t_0)} > 0.$$

This contradicts $t_0$ being a point of maximum in $[a_1, a_2]$ and hence $p(t) \leq x(t)$ on $[a_1, a_2]$. By a similar argument $x(t) \leq q(t)$ on $[a_1, a_2]$ and therefore $p(t) \leq x(t) \leq q(t)$ on $[a_1, a_2]$.

**Lemma 7.5 [8]** Suppose solutions of initial value problems for (7.1) extend to $[a_1, a_2]$ or become unbounded. If $\{x_m(t)\}$ is a uniformly bounded sequence of solutions of $x'' = f_m(t, x, x')$, where $\{f_m\}$ is a sequence of functions defined and continuous on $[a_1, a_2] \times R^2$ which converges uniformly to f on compact subsets of $[a_1, a_2] \times R^2$, then there is a solution $x(t)$ of (7.1) defined on $[a_1, a_2]$ and a subsequence of $\{x_m(t)\}$ which converges uniformly to $x(t)$ on $[a_1, a_2]$.

**Proof.** Since $|x_m(t)| \leq M$ on $[a_1, a_2]$ for some M and $m \geq 1$, we have

$$\frac{2M}{a_2 - a_1} \geq \left| \frac{x_m(a_2) - x_m(a_1)}{a_2 - a_1} \right| = |x'_m(\xi_m)|$$

where $a_1 < \xi_m < a_2$. Consequently, $\{\xi_m\}$, $\{x_m(\xi_m)\}$ and $\{x'_m(\xi_m)\}$ are bounded sequences. By taking subsequences in succession which converge, we conclude that there exist values $\xi_0$, $x_0$, $x'_0$ such that $\xi_{m(1)} \to \xi_0$, $x_{m(1)}(\xi_{m(1)}) \to x_0$, $x'_{m(1)}(\xi_{m(1)}) \to x'_0$, where $\{m(1)\}$ is some subsequence of $\{m\}$. Thus by the standard convergence theorem [6, p.14] there is a subsequence $\{x_{m(2)}(t)\}$ of $\{x_{m(1)}(t)\}$ and a solution $x(t)$ of (7.1) which is the uniform limit of $\{x_{m(2)}(t)\}$ on compact subintervals of the interval of existence of $x(t)$. This $x(t)$ cannot become unbounded because $\{x_{m(2)}(t)\}$ is uniformly bounded, hence $x(t)$ is defined on $[a_1, a_2]$ and convergence is uniform on $[a_1, a_2]$.

**Theorem 7.6 [1]** Let $f \in C[[a_1, a_2] \times R^2, R]$ and suppose that solutions of initial value problems for (7.1) extend to $[a_1, a_2]$ or become unbounded.

Then, a necessary and sufficient condition that the BVP (7.1), (7.7),
(7.8) has a solution is that there exist lower and upper solutions p
and q of (7.1), (7.7), (7.8) satisfying $p(t) \leq q(t)$ on $[a_1, a_2]$. In the
sufficiency part, the solution $x(t)$ satisfies $p(t) \leq x(t) \leq q(t)$.

Proof. The necessity is easily verified by choosing $p(t) = q(t) = x(t)$
on $[a_1, a_2]$, where $x(t)$ is the assumed solution of (7.1), (7.7), (7.8).

To prove the sufficiency part the function f is truncated in the
following manner. Let

$$f_m(t,x,x') = f(t,x,m) \quad \text{if} \quad x' > m$$

$$= f(t,x,x') \quad \text{if} \quad |x'| \leq m$$

$$= f(t,x,-m) \quad \text{if} \quad x' < -m$$

where m is any integer satisfying $m \geq N_0$ and

$$N_0 = \max \left\{ \max_{a_1 \leq t \leq a_2} |p'(t)|, \quad \max_{a_1 \leq t \leq a_2} |q'(t)| \right\}.$$

Also, let

$$F_m(t,x,x') = f_m(t,q(t),x') + \frac{x - q(t)}{1 + x^2} \quad \text{if} \quad x > q(t)$$

$$= f_m(t,x,x') \quad \text{if} \quad p(t) \leq x \leq q(t)$$

$$= f_m(t,p(t),x') - \frac{p(t) - x}{1 + x^2} \quad \text{if} \quad x < p(t).$$

Then, $F_m$ is continuous and bounded on $[a_1, a_2] \times R^2$ for each $m \geq N_0$.
Hence, by Corollary 7.3 there exists a solution $x_m(t)$ of $x'' = F_m(t,x,x')$
which satisfies the boundary conditions (7.7), (7.8). Also, as in
Theorem 7.4 this $x_m(t)$ will satisfy $p(t) \leq x_m(t) \leq q(t)$ on $[a_1, a_2]$, and
hence $x_m(t)$ is a solution of $x'' = f_m(t,x,x')$ satisfying the boundary

conditions (7.7), (7.8). Consequently for $m \geq N_0$, $\{x_m(t)\}$ is a sequence of solutions of $x'' = f_m(t,x,x')$, (7.7), (7.8) which is uniformly bounded on $[a_1,a_2]$. Moreover, $f_m$ converges to $f$ uniformly on compact subsets of $[a_1,a_2] \times R^2$. The conditions of Lemma 7.5 are satisfied and thus a subsequence of $\{x_m(t)\}$ converges uniformly to a solution of (7.1), (7.7), (7.8) satisfying $p(t) \leq x(t) \leq q(t)$ on $[a_1,a_2]$.

**Definition 7.2**    Let $f \in C[[a_1,a_2] \times R^2, R]$ and $p, q \in C^{(2)}[a_1,a_2]$ with $p(t) \leq q(t)$ on $[a_1,a_2]$. Suppose that for $t \in [a_1,a_2]$, $p(t) \leq x_0 \leq q(t)$ and $x_1 \in R$, $|f(t,x_0,x_1)| \leq h(|x_1|)$ where $h \in C[R_+, (0,\infty)]$. If

(7.17)
$$\int_\lambda^\infty \frac{s \, ds}{h(s)} > \max_{a_1 \leq t \leq a_2} q(t) - \min_{a_1 \leq t \leq a_2} p(t)$$

where

(7.18)
$$\lambda(a_2-a_1) = \max\{|p(a_1)-q(a_2)|, |p(a_2)-q(a_1)|\}$$

we say that $f$ satisfies Nagumo's condition on $[a_1,a_2]$ relative to $p,q$.

**Lemma 7.7 [7]**    Assume that $f$ satisfies Nagumo's condition on $[a_1,a_2]$ with respect to the pair $p$, $q$. Then, for any solution $x(t)$ of (7.1) with $p(t) \leq x(t) \leq q(t)$ on $[a_1,a_2]$ there exists an $N > 0$ depending only on $p$, $q$, $h$ such that

(7.19)
$$|x'(t)| \leq N \quad \text{on } [a_1,a_2].$$

**Proof.**    Because of (7.17), we can choose an $N > \lambda$ such that

$$\int_\lambda^N \frac{s \, ds}{h(s)} > \max_{a_1 \leq t \leq a_2} q(t) - \min_{a_1 \leq t \leq a_2} p(t).$$

If $t_0 \in (a_1,a_2)$ is such that $(a_2-a_1)x'(t_0) = x(a_2) - x(a_1)$, then by (7.18) we have $|x'(t_0)| \leq \lambda$. Assume that (7.19) is not true. Then, there exists an interval $[t_1,t_2] \subset [a_1,a_2]$ such that the following cases hold :

(i)   $x'(t_1) = \lambda$, $x'(t_2) = N$ and $\lambda < x'(t) < N$, $t \in (t_1,t_2)$

(ii)  $x'(t_1) = N$, $x'(t_2) = \lambda$ and $\lambda < x'(t) < N$, $t \in (t_1,t_2)$

(iii) $x'(t_1) = -\lambda$, $x'(t_2) = -N$ and $-N < x'(t) < -\lambda$, $t \in (t_1,t_2)$

(iv)  $x'(t_1) = -N$, $x'(t_2) = -\lambda$ and $-N < x'(t) < -\lambda$, $t \in (t_1,t_2)$.

Let us consider case (i).  On $[t_1,t_2]$, we have

$$|x''(t)|\,|x'(t)| = |f(t,x(t),x'(t))|\,|x'(t)| \le h(x'(t))x'(t)$$

and hence

$$\left|\int_{t_1}^{t_2} \frac{x''(s)x'(s)ds}{h(x'(s))}\right| \le \int_{t_1}^{t_2} \frac{|x''(s)|\,|x'(s)|ds}{h(x'(s))} \le \int_{t_1}^{t_2} x'(s)ds.$$

This leads to the contradiction

$$\int_{\lambda}^{N} \frac{s\,ds}{h(s)} \le x(t_2) - x(t_1) \le \max_{a_1 \le t \le a_2} q(t) - \min_{a_1 \le t \le a_2} p(t).$$

We can deal with the remaining possibilities in a similar way and therefore we conclude that (7.19) is valid.

Corollary 7.8   Let $f \in C[[a_1,a_2] \times R^2, R]$ and $p$, $q \in C^{(2)}[a_1,a_2]$ with
$p(t) \le q(t)$ on $[a_1,a_2]$.  Suppose that for $t \in [a_1,a_2]$, $p(t) \le x_0 \le q(t)$
and $x_1$, $y_1 \in R$

(7.20)                $|f(t,x_0,x_1) - f(t,x_0,y_1)| \le L\,|x_1 - y_1|$

where $L > 0$ is a constant.  Then, $f$ satisfies Nagumo's condition on
$[a_1,a_2]$ with respect to the pair $p,q$.

Theorem 7.9   Let $f \in C[[a_1,a_2] \times R^2, R]$ and satisfy  Nagumo's  condition
with respect to the pair $p,q$ which are lower and upper solutions of the

BVP (7.1), (7.7), (7.8) respectively. Then, there exists a solution
$x(t)$ of the BVP (7.1), (7.7), (7.8) such that $p(t) \leq x(t) \leq q(t)$ on
$[a_1,a_2]$.

Proof. By Lemma 7.7 there is an $N > 0$ depending only on $p,q$ and $h$
such that $|x'(t)| \leq N$ on $[a_1,a_2]$ for any solution $x(t)$ of (7.1) with
$p(t) \leq x(t) \leq q(t)$ on $[a_1,a_2]$. Choose a $c_1 > N$ so that $|p'(t)| < c_1$,
$|q'(t)| < c_1$ on $[a_1,a_2]$. Now, we define

$$F^*(t,x,x') = f(t,x,c_1) \quad \text{if } x' \geq c_1$$
$$= f(t,x,x') \quad \text{if } |x'| \leq c_1$$
$$= f(t,x,-c_1) \quad \text{if } x' \leq -c_1$$

and

$$F(t,x,x') = F^*(t,q(t),x') + \frac{x-q(t)}{1+x^2} \quad \text{if } x > q(t)$$
$$= F^*(t,x,x') \quad \text{if } p(t) \leq x \leq q(t)$$
$$= F^*(t,p(t),x') - \frac{p(t)-x}{1+x^2} \quad \text{if } x < p(t).$$

The function $F$ so defined is continuous on $[a_1,a_2] \times R^2$ and satisfy
the conditions of Corollary 7.3. Hence, there exists a solution $x(t)$
of

$$x'' = F(t,x,x')$$

which satisfies the boundary conditions (7.7), (7.8). Also, as in
Theorem 7.4 this solution is such that $p(t) \leq x(t) \leq q(t)$ on $[a_1,a_2]$.
By the mean-value theorem there is a $t_0 \in (a_1,a_2)$ such that

$$(a_2-a_1)x'(t_0) = x(a_2) - x(a_1)$$

and it follows that $|x'(t_0)| \leq \lambda < N < c_1$. Thus, there is an interval
around $t_0$ in which $x(t)$ is a solution of (7.1). By Lemma 7.7, we have
$|x'(t)| \leq N < c_1$ in this interval. However, $x(t)$ is a solution of (7.1)
as long as $|x'(t)| < c_1$. We conclude that $x(t)$ is a solution of (7.1),

72

(7.7), (7.8) on $[a_1,a_2]$.

**Corollary 7.10 [13]**  Let the conditions of Corollary 7.8 be satisfied, and p,q be the lower and upper solutions of the BVP (7.1), (7.7), (7.8). Then, there exists a solution x(t) of the BVP (7.1), (7.7), (7.8) such that $p(t) \le x(t) \le q(t)$ on $[a_1,a_2]$.

**Theorem 7.11**  Let $f \in C[[a_1,a_2] \times R^2, R]$ and $f(t,x_0,x_1)$ be strictly increasing in $x_0$ for each $(t,x_1) \in [a_1,a_2] \times R$. Further, we assume that there exist p,q lower and upper solutions of (7.1), (7.7), (7.8). Then, $p(t) \le q(t)$ on $[a_1,a_2]$.

**Proof.**  The proof is similar to that of Theorem 7.4.

**Corollary 7.12**  Let $f \in C[[a_1,a_2] \times R^2, R]$ and $f(t,x_0,x_1)$ be strictly increasing in $x_0$ for each $(t,x_1) \in [a_1,a_2] \times R$. Then, the BVP (7.1), (7.7), (7.8) has at most one solution.

**Theorem 7.13 [4]**  Let $f \in C[[a_1,a_2] \times R^2, R]$ and $f(t,x_0,x_1)$ be non-decreasing in $x_0$ for each $(t,x_1) \in [a_1,a_2] \times R$. Further, let for all $(t,x_0,x_1)$, $(t,x_0,y_1) \in [a_1,a_2] \times R^2$ the inequality (7.20) be satisfied. Then, the BVP (7.1), (7.7), (7.8) has a unique solution.

**Proof.**  We shall prove only the existence part. For this, let $m = \min\limits_{a_1 \le t \le a_2} f(t,0,0)$ and $M = \max\limits_{a_1 \le t \le a_2} f(t,0,0)$. Consider the differential equations

(7.21)  $$x'' + L|x'| - m = 0$$
and
(7.22)  $$x'' - L|x'| - M = 0.$$

Let q(t) be a nonnegative solution of (7.21) which satisfies (7.15) and (7.16), and p(t) be a nonpositive solution of (7.22) which satisfies (7.12) and (7.13), such solutions may be computed explicitly.   Then, we have

$$f(t,q(t),q'(t)) \geq f(t,0,q'(t)) \geq - L|q'(t)| + f(t,0,0)$$
$$\geq - L|q'(t)| + m = q''(t)$$

and

$$f(t,p(t),p'(t)) \leq f(t,0,p'(t)) \leq L|p'(t)| + f(t,0,0)$$
$$\leq L|p'(t)| + M = p''(t).$$

Thus, p and q are lower and upper solutions of (7.1), (7.7), (7.8). Hence, from Corollary 7.10 there exists a solution x(t) of the BVP (7.1), (7.7), (7.8) such that $p(t) \leq x(t) \leq q(t)$.

**Theorem 7.14** [1]  Let f e $[[a_1,a_2] \times R^2, R]$ and satisfy the Lipschitz condition (7.3).  Then, if $K_1$ and $K_2$ defined in Theorem 7.2 are such that

(7.23) $$K_1 L_0 + K_2 L_1 < 1$$

the BVP (7.1), (7.7), (7.8) has a unique solution.

**Proof.**  The proof consists of a standard application of the Contraction Mapping Principle.

In particular, if $\alpha_0 = \beta_0 = 1$ and $\alpha_1 = \beta_1 = 0$ then (7.23) reduces to (7.4).

In the x - x' plane the boundary conditions (7.7), (7.8) represent straight lines, (7.7) represents an initial line and (7.8) a terminal line.  We shall represent these lines as $\ell_1(a_1) = A$ and $\ell_2(a_2) = B$.

Consider

(7.24) $$x'' + Mx' + K x = 0$$

(7.25) $$\alpha_0 x(0) - \alpha_1 x'(0) = 0 \text{ or } \ell = 0.$$

Define $\alpha(M,K,\ell)$ and $-\beta(M,K,\ell)$ as the t-values of the next and of the preceding zero of $x'(t)$ for a solution $x(t)$ of (7.24) and (7.25) if such exist, and $+\infty$ and $-\infty$, respectively, otherwise. If $\ell = 0$ is the x axis, both $\alpha(M,K,\ell)$ and $\beta(M,K,\ell)$ are taken to be zero. In the $x-x'$ plane this is just time to traverse the angle between the line $\ell = 0$ and the x axis. Since the equation is linear $\alpha(M,K,\ell)$ and $\beta(M,K,\ell)$ are independent of the initial position on $\ell = 0$ and since the equation has constant coefficients these quantities can be computed explicitly and are independent of the starting time $t = 0$.

**Theorem 7.15 [14]**   Let $f \in C[[a_1,a_2] \times R^2, R]$ and

$$M_1(y_1 - x_1) \le f(t,x_0,x_1) - f(t,x_0,y_1) \le M_2(y_1 - x_1), \; y_1 \ge x_1$$

$$f(t,x_0,x_1) - f(t,y_0,x_1) \le K(y_0-x_0), \; y_0 \ge x_0.$$

Further, let solutions of the initial value problem for (1.1) at $t = a_1$ exist on $[a_1,a_2]$ and be unique. Then, if

(7.26)      $a_2 - a_1 < \alpha(M_2,K,\ell_1) + \beta(M_1,K,\ell_2)$

the BVP (7.1), (7.7), (7.8) has a unique solution. This result is best possible.

Obviously, the conclusions of Theorem 7.1 are included in Theorem 7.15. Further, since there is no sign condition on the constants $M_1$, $M_2$ and K this result is more useful in applications. In fact from Theorem 7.15 the BVP : $x'' - x = 0$, (7.2) has a unique solution on all finite intervals $[a_1,a_2]$ whereas Theorem 7.1 requires $a_2 - a_1 < \pi$.

# COMMENTS AND BIBLIOGRAPHY

Theory of differential inequalities furnishes a very general comparison principle for studying many qualitative as well as quantitative properties of solutions of differential equations. In particular, for the two point BVPs this theory has been developed and applied fruitfully for the existence and uniqueness of the solutions [5,7].

1. Agarwal, R. P. and Srivastava, U. N. "Generalized two-point boundary value problems", J. Math. Phyl. Sci. 10, 367-373 (1976).
2. Bailey, P. and Waltman, P. "On the distance between consecutive zeros for second order differential equations", J. Math. Anal. Appl. 14, 23-30 (1966).
3. Bailey, P. B., Shampine, L. F. and Waltman, P. E. Nonlinear Two Point Boundary Value Problems, Academic Press, New York, 1968.
4. Bebernes, J. W. and Gaines, R. "A generalized two-point boundary value problem", Proc. Amer. Math. Soc. 19, 749-754 (1968).
5. Bernfeld, S. R. and Lakshmikantham, V. An Introduction to Nonlinear Boundary Value Problems, Academic Press. New York, 1974.
6. Hartman, P. Ordinary Differential Equations, Wiley, New York, 1964.
7. Jackson, L. K. "Subfunctions and second-order ordinary differential inequalities", Advances in Math. 2, 307-363 (1968).
8. Klaasen, G. A. "Differential inequalities and existence theorems for second and third order boundary value problems", J. Diff. Equs. 10, 529-537 (1971).
9. Lettenmeyer, F. "Über die von einem Punkt ausgehenden Integralkurven einer Differentialgleichung 2. Ordnung", Deutsche Math. 7, 56-74 (1944).
10. Opial, Z. "Sur une inégalité de C. de la Vallée Poussin dans la théorie de l'équation différentielle linéaire du second order", Ann. Polon. Math. 6, 87-91 (1959).
11. Picard, E. "Sur l'application des méthodes d'approximations successives á l'étude de certaines équations différentielles ordinaires", J. Math. 9, 217-271 (1893).
12. Picard, E. Traité d'Analyse, Vol. III. Paris, 1908.
13. Schmitt, K. "A nonlinear boundary value problem", J. Diff. Equs. 9, 527-537 (1970).
14. Waltman, P. "A nonlinear boundary value problem", J. Diff. Equs. 4, 597-603 (1968).

## 8. ERROR ESTIMATES IN POLYNOMIAL INTERPOLATION

In polynomial interpolation theory, the following result is well known :

Theorem 8.1    Let $x(t) \in C^{(n)}[a,b]$, satisfying (3.8).  Then,

$$(8.1) \qquad |x^{(k)}(t)| \leq \alpha_{n,k} \, \mu(b-a)^{n-k}, \; 0 \leq k \leq n-1$$

where $\mu = \max\limits_{a \leq t \leq b} |x^{(n)}(t)|$, and

$$\alpha_{n,k} = \frac{1}{(n-k)!} \, .$$

The proof follows from the Hermite interpolation formula

$$x(t) = \frac{1}{n!} \, P(t) x^{(n)}(\xi)$$

where $\xi \in (a,b)$, and from the observation that the kth derivative of $x(t)$ has at least n-k zeros in (a,b).  The constants $\alpha_{n,k}$ in (8.1) obviously are the best possible.

If we consider only the segment $[a_1, a_r]$, which corresponds to interpolation in the exact sense of the word, then the inequalities (8.1) can be improved :

Theorem 8.2    Let $x(t) \in C^{(n)}[a_1, a_r]$, satisfying (3.8).  Then,

$$(8.2) \qquad |x^{(k)}(t)| \leq \beta_{n,k} \, m(a_r - a_1)^{n-k}, \; 0 \leq k \leq n-1$$

where $m = \max\limits_{a_1 \leq t \leq a_r} |x^{(n)}(t)|$, and

$$\beta_{n,0} = \frac{1}{n!} \frac{(n-1)^{n-1}}{n^n} , \quad \beta_{n,k} = \frac{k}{n(n-k)!} , \quad 1 \le k \le n-1.$$

This theorem has been proved in two different ways, first using a theorem due to Krein and Milman concerning extremal points and the second using a suitable integral representation of $x(t)$. Hukuhara [9] indicates that Tumura [12] proved this result, and it has been used in [4, 6]. In (8.2) the constants $\beta_{n,k}$, $0 \le k \le n-1$ are the best possible, as they are exact for the functions

$$x_1(t) = (t-a_1)^{n-1}(a_r-t), \quad x_2(t) = (t-a_1)(a_r-t)^{n-1}$$

and only for these functions, upto a constant factor.

The constants $\beta_{n,k}$ are free from any nature of multiplicity at the points $a_i$, $1 \le i \le r$. For $\alpha = \min(k_1,k_r)$, we shall establish the following result :

Theorem 8.3 [1]    Let $x(t) \in C^{(n)}[a_1,a_r]$, satisfying (3.8) and $\alpha = \min(k_1,k_r)$. Then,

$$(8.3) \qquad |x^{(k)}(t)| \le C_{n,k} \, m(a_r-a_1)^{n-k}, \quad 0 \le k \le n-1$$

where m is same as in Theorem 8.2, and

$$C_{n,k} = \frac{1}{(n-k)!} \frac{(n-\alpha-1)^{n-\alpha-1}(\alpha-k+1)^{\alpha-k+1}}{(n-k)^{n-k}} , \quad 0 \le k \le \alpha$$

$$C_{n,k+\alpha} = \frac{k}{(n-\alpha)(n-\alpha-k)!} , \quad 1 \le k \le n-\alpha-1.$$

The constants $C_{n,k}$ are smaller than the corresponding $\beta_{n,k}$.

Proof.  First, we shall prove the theorem for $0 \le k \le \alpha$. Since $k_1+1$

and $k_r+1$-are the multiplicities of the zeros at $a_1$ and $a_r$ respectively, from Rolle's theorem it follows that $x^{(k)}(t)$ has at least $(n-k_1-k_r+k-2)$ $= N$ (say) zeros in $(a_1,a_r)$, $(k_1-k+1)$ at $a_1$ and $(k_r-k+1)$ at $a_r$. Define, $h(t) = x^{(k)}(t)$, then

$$h(a_1) = h'(a_1) = \ldots = h^{(k_1-k)}(a_1) = 0$$

$$h(a_{k,i}) = h'(a_{k,i}) = \ldots = h^{(p_i)}(a_{k,i}) = 0, \; a_{k,i} \in (a_1,a_r)$$

(8.4)

$$p_i \geq 0, \; 1 \leq i \leq q; \; a_{k,j} < a_{k,j+1}, \; 1 \leq j \leq q-1, \; \sum_{i=1}^{q} p_i + q = N$$

$$h(a_r) = h'(a_r) = \ldots = h^{(k_r-k)}(a_r) = 0.$$

Let $g_k(t,s)$ be the Green's function of the BVP : $h^{(n-k)}(t) = 0$, satisfying (8.4). Then, $h(t)$ can be written as

$$h(t) = \int_{a_1}^{a_r} g_k(t,s)h^{(n-k)}(s)ds.$$

Hence,

$$|h(t)| \leq \max_{a_1 \leq t \leq a_r} |h^{(n-k)}(t)| \int_{a_1}^{a_r} |g_k(t,s)|ds.$$

Now an application of Lemma 3.3 provides

(8.5)    $$|h(t)| \leq m \frac{1}{(n-k)!} Q_k(t)$$

where

$$Q_k(t) = (t-a_1)^{k_1-k+1}(a_r-t)^{k_r-k+1} \prod_{i=1}^{q} |t-a_{k,i}|^{p_i+1}.$$

Next, suppose that $a_{k,j} \leq t \leq a_{k,j+1}$, $0 \leq j \leq q$ where $a_{k,0} = a_1$ and $a_{k,q+1} = a_r$ then, it is easy to verify that

$$Q_k(t) \leq (t-a_1)^{k_1-k+j+1}(a_r-t)^{n-k_1-j-1}$$

$$\leq \begin{cases} (t-a_1)^{n-\alpha-1}(a_r-t)^{\alpha-k+1} = \phi_k(t), & t-a_1 \geq a_r-t \\ (t-a_1)^{\alpha-k+1}(a_r-t)^{n-\alpha-1} = \psi_k(t), & t-a_1 \leq a_r-t. \end{cases}$$

The absolute maximum of $\phi_k(t)$ occurs at $t = a_1 + \frac{n-\alpha-1}{n-k}(a_r-a_1)$ and of $\psi_k(t)$ at $t = a_1 + \frac{\alpha-k+1}{n-k}(a_r-a_1)$. Also, the absolute maximum value of $\phi_k(t)$ or $\psi_k(t)$ is the same, namely $\frac{(n-\alpha-1)^{n-\alpha-1}(\alpha-k+1)^{\alpha-k+1}}{(n-k)^{n-k}}(a_r-a_1)^{n-k}$.

Thus, inequality (8.3) for $0 \leq k \leq \alpha$ follows from (8.5).

To show that the $C_{n,k}$ are smaller than the corresponding $\beta_{n,k}$ for $0 \leq k \leq \alpha$, we note that

$$\phi_k(t) \leq \frac{n-k}{n}(t-a_1)^{n-k-1}(a_r-t) + \frac{k}{n}(t-a_1)^{n-k} = \phi_k^*(t)$$

$$\psi_k(t) \leq \frac{k}{n}(a_r-t)^{n-k} + \frac{n-k}{n}(t-a_1)(a_r-t)^{n-k-1} = \psi_k^*(t), \quad 0 \leq k \leq \alpha.$$

For $k = 0$, $\phi_k^*(t)$ has an absolute maximum at $t = \frac{(n-1)a_r+a_1}{n}$ and $\psi_k^*(t)$ at $t = \frac{(n-1)a_1+a_r}{n}$. Also, for both the absolute maximum value is $\frac{(n-1)^{n-1}}{n^n}(a_r-a_1)^n$. For $1 \leq k \leq \alpha$, $\phi_k^*(t)$ has an absolute maximum at $t = a_r$ and $\psi_k^*(t)$ at $t = a_1$. Also, for both the absolute maximum value is $\frac{k}{n}(a_r-a_1)^{n-k}$.

Next, to prove Theorem 8.3 for $\alpha+1 \leq k \leq n-1$, we observe that $x^{(\alpha)}(t)$ has one zero at $a_1$ (or at $a_r$), and that at $a_r$ (or at $a_1$) there may be more than one zero, also at least $n - k_1 - k_r + \alpha - 2$ zeros in $(a_1, a_r)$.

Define, $h(t) = x^{(\alpha)}(t)$ then, from Theorem 8.2 we obtain

$$\left| h^{(k)}(t) \right| \leq \frac{k}{(n-\alpha)(n-\alpha-k)!} \, m \, (a_r-a_1)^{n-\alpha-k}, \; 1 \leq k \leq n-\alpha-1$$

which proves (8.3) for $\alpha+1 \leq k \leq n-1$. Since it is easy to verify that $C_{n,k+\alpha} \leq \beta_{n,k+\alpha}$, $1 \leq k \leq n-\alpha-1$ the proof of Theorem 8.3 is complete.

The constant $C_{n,0}$ is the best possible, as it is exact for the functions

$$x_1(t) = (t-a_1)^{n-\alpha-1}(a_r-t)^{\alpha+1}, \quad x_2(t) = (t-a_1)^{\alpha+1}(a_r-t)^{n-\alpha-1}$$

and only for these functions, upto a constant factor. Finding the best possible constants $C_{n,k}$ for the case $k \neq 0$, $\alpha \neq 0$ is an <u>open problem</u>.

For the particular case : $n = 4$, $\alpha = 1$, i.e., $x(a_1) = x'(a_1) = x(a_2) = x'(a_2) = 0$ the comparison between $\beta_{n,k}$, $C_{n,k}$ and the best possible constants obtained in [5] is the following

|       | $\beta_{4,k}$ | $C_{4,k}$ | Best Possible |
|-------|---------------|-----------|---------------|
| k=0   | $\dfrac{9}{2048}$ | $\dfrac{1}{384}$ | $\dfrac{1}{384}$ |
| k=1   | $\dfrac{1}{24}$ | $\dfrac{2}{81}$ | $\dfrac{1}{72\sqrt{3}}$ |
| k=2   | $\dfrac{1}{4}$ | $\dfrac{1}{6}$ | $\dfrac{1}{12}$ |
| k=3   | $\dfrac{3}{4}$ | $\dfrac{2}{3}$ | $\dfrac{1}{2}$ . |

In the same paper [5] the best possible constants $C_{n,k}$ for the case $n = 6$, $\alpha = 2$, i.e., $x(a_1) = x'(a_1) = x''(a_1) = x(a_2) = x'(a_2) = x''(a_2)=0$

are also given and appear as $C_{6,0} = \frac{1}{46080}$, $C_{6,1} = \frac{\sqrt{5}}{30000}$, $C_{6,2} = \frac{1}{1920}$, $C_{6,3} = \frac{1}{120}$, $C_{6,4} = \frac{1}{10}$, $C_{6,5} = \frac{1}{2}$.

**Theorem 8.4** [7, 10]  Let $x(t) \in C^{(n)}[a,b]$, satisfying

$$(8.6) \qquad x^{(i)}(a_{i+1}) = 0, \quad 0 \le i \le n-1$$

where $a \le a_1 \le a_2 \le \cdots \le a_n \le b$.  Then,

$$(8.7) \qquad |x(t)| \le \mu \frac{(b-a)^n}{n[\frac{n-1}{2}]! \, [\frac{n}{2}]!}$$

where $\mu$ is same as in Theorem 8.1.

**Proof.**  From the identity

$$x^{(k)}(t) = \int_{a_{k+1}}^{t} \int_{a_{k+2}}^{t_1} \cdots \int_{a_n}^{t_{n-k-1}} x^{(n)}(t_{n-k}) dt_{n-k} \, dt_{n-k-1} \cdots dt_1$$

for $t \le a_{k+1}$, we find

$$|x^{(k)}(t)| \le \int_{t}^{b} \int_{t_1}^{b} \cdots \int_{t_{n-k-1}}^{b} |x^{(n)}(t_{n-k})| dt_{n-k} \, dt_{n-k-1} \cdots dt_1$$

$$(8.8) \qquad \le \mu \frac{(b-t)^{n-k}}{(n-k)!} \, .$$

In particular

$$|x(t)| \le \mu \frac{(b-a)^n}{n!} \quad \text{for } a \le t \le a_1.$$

Similarly, from

$$x(t) = \int_{a_1}^{t} \int_{a_2}^{t_1} \cdots \int_{a_k}^{t_{k-1}} x^{(k)}(t_k) dt_k \, dt_{k-1} \cdots dt_1$$

for $t \geq a_k$, we obtain

$$|x(t)| \leq \int_{a}^{t} \int_{a}^{t_1} \cdots \int_{a}^{t_{k-1}} |x^{(k)}(t_k)| dt_k \, dt_{k-1} \cdots dt_1$$

(8.9)
$$= \int_{a}^{t} \frac{(t-s)^{k-1}}{(k-1)!} |x^{(k)}(s)| ds.$$

In particular

$$|x(t)| \leq \mu \int_{a}^{b} \frac{(b-s)^{n-1}}{(n-1)!} ds = \mu \frac{(b-a)^{n}}{n!} \text{ for } a_n \leq t \leq b.$$

From (8.8) and (8.9) it follows that for $a_k \leq t \leq a_{k+1}$

$$|x(t)| \leq \mu \int_{a}^{t} \frac{(b-s)^{n-k}}{(n-k)!} \frac{(t-s)^{k-1}}{(k-1)!} ds$$

$$\leq \mu \int_{a}^{b} \frac{(b-s)^{n-1}}{(n-k)!(k-1)!} ds$$

$$= \mu \frac{(b-a)^{n}}{n(n-k)!(k-1)!}$$

$$= \mu \binom{n-1}{n-k} \frac{(b-a)^{n}}{n!}.$$

When $k$ runs through the values $1,\ldots,n-1$ the Binomial coefficient $\binom{n-1}{n-k}$ takes its maximum value for $k = [\frac{n+1}{2}]$ and hence (8.7) follows.

From the proof it is clear that in (8.7) equality cannot hold unless x(t) is a polynomial of degree n.

**Theorem 8.5** [3] Let $x(t) \in C^{(n)}[a_1, a_2]$, satisfying (3.15). Then,

$$(8.10) \qquad |x^{(i)}(t)| \leq D_{n,i} \, m(a_2 - a_1)^{n-i}, \quad 0 \leq i \leq n-1$$

where $m = \max\limits_{a_1 \leq t \leq a_2} |x^{(n)}(t)|$, and

$$D_{n,i} = \frac{1}{(n-i)!} \left| \sum_{j=0}^{k-i-1} \binom{n-i}{j}(-1)^{n-i-j} \right|, \quad 0 \leq i \leq k-1$$

$$= \frac{1}{(n-i)!} \, , \quad k \leq i \leq n-1.$$

**Proof.** Any such function can be written as

$$x(t) = \int_{a_1}^{a_2} g_4(t,s) x^{(n)}(s) \, ds$$

and hence

$$|x^{(i)}(t)| \leq \left( \max_{a_1 \leq t \leq a_2} \int_{a_1}^{a_2} |g_4^{(i)}(t,s)| ds \right) \max_{a_1 \leq t \leq a_2} |x^{(n)}(t)|.$$

Thus, it suffices to show that

$$\max_{a_1 \leq t \leq a_2} \int_{a_1}^{a_2} |g_4^{(i)}(t,s)| ds \leq D_{n,i}(a_2 - a_1)^{n-i}, \quad 0 \leq i \leq n-1.$$

From (3.16) and (3.17) for $0 \leq i \leq k-1$, we have

$$\max_{a_1 \le t \le a_2} \int_{a_1}^{a_2} |g_4^{(i)}(t,s)| ds$$

$$= \max_{a_1 \le t \le a_2} \frac{1}{(n-1)!} \left| - \sum_{j=i}^{k-1} \binom{n-1}{j} \frac{j!}{(j-i)!} (t-a_1)^{j-i} \frac{(a_1-t)^{n-j}}{n-j} \right.$$

$$\left. + \sum_{j=k}^{n-1} \binom{n-1}{j} \frac{j!}{(j-i)!} (t-a_1)^{j-i} \frac{(a_1-a_2)^{n-j} - (a_1-t)^{n-j}}{n-j} \right|$$

$$= \max_{a_1 \le t \le a_2} \left| - \sum_{j=i}^{n-1} \frac{(-1)^{n-j}}{(n-j)!(j-i)!} (t-a_1)^{n-i} \right.$$

$$\left. + \sum_{j=k}^{n-1} \frac{(-1)^{n-j}}{(n-j)!(j-i)!} (a_2-a_1)^{n-j} (t-a_1)^{j-i} \right|$$

$$= \frac{1}{(n-i)!} \max_{a_1 \le t \le a_2} \left| (t-a_1)^{n-i} + \sum_{j=k-i}^{n-i-1} \binom{n-i}{j} (a_1-a_2)^{n-i-j} (t-a_1)^j \right|$$

$$= \frac{1}{(n-i)!} \max_{a_1 \le t \le a_2} \left| \sum_{j=k-i}^{n-i} \binom{n-i}{j} (a_1-a_2)^{n-i-j} (t-a_1)^j \right|$$

$$= \frac{1}{(n-i)!} \left| \sum_{j=k-i}^{n-i} \binom{n-i}{j} (-1)^{n-i-j} \right| (a_2-a_1)^{n-i}$$

$$= D_{n,i} (a_2-a_1)^{n-i}.$$

Similarly, from (3.18) and (3.19) for $k \le i \le n-1$, we have

$$\max_{a_1 \le t \le a_2} \int_{a_1}^{a_2} |g_4^{(i)}(t,s)| ds = \max_{a_1 \le t \le a_2} \frac{1}{(n-i-1)!} \int_{t}^{a_2} (s-t)^{n-i-1} ds$$

$$= \frac{1}{(n-i)!} (a_2-a_1)^{n-i}.$$

In (8.10) the constants $D_{n,i}$ are the best possible as they are exact for the function

$$x(t) = \frac{1}{n!} \sum_{i=k}^{n} \binom{n}{i} (a_1-a_2)^{n-i} (t-a_1)^i$$

and only for this function upto a constant factor.

**Theorem 8.6 [2]** Let $x(t) \in C^{(n)}[a_1,a_2]$, satisfying (3.20) or (3.24). Then,

(8.11)  $\qquad |x^{(k)}(t)| \leq E_{n,k} \, m(a_2-a_1)^{n-k}, \; 0 \leq k \leq n-1$

where m is same as in theorem 8.5, and

$$E_{n,k} = \begin{cases} \dfrac{(n-k-1)^{n-k-1}}{(n-k)! \, (n-p)^{n-k}} & n-1 \geq p = k \\[3ex] \dfrac{(p-k)}{(n-p)(n-k)!} & n-1 \geq p \geq k+1 \end{cases}$$

$$E_{n,p+k} = \frac{k}{(n-p)(n-p-k)!} \qquad 1 \leq k \leq n-p-1.$$

**Proof.** Any function satisfying (3.20) can be written as

$$x(t) = \int_{a_1}^{a_2} g_5(t,s) x^{(n)}(s) ds$$

and hence from (3.22) for $0 \leq k \leq p$, we find

$$|x^{(k)}(t)| \leq m \int_{a_1}^{a_2} [-g_5^{(k)}(t,s)] ds$$

$$= m \frac{1}{(n-k-1)!} (t-a_1)^{n-k-1} [\frac{a_2-a_1}{n-p} - \frac{t-a_1}{n-k}]$$

$$= m\phi_k(t).$$

Similarly, from Lemma 3.7 for any function satisfying (3.24), we get

$$|x^{(k)}(t)| \le m \frac{1}{(n-k-1)!} (a_2-t)^{n-k-1} [\frac{a_2-a_1}{n-p} - \frac{a_2-t}{n-k}]$$

$$= m\psi_k(t).$$

The function $\phi_k(t)$ attains an absolute maximum at $t = a_1 + \frac{(n-k-1)}{(n-p)}(a_2-a_1)$ if $k = p \le n-1$ and at $t = a_2$ if $k+1 \le p \le n-1$. Also $\psi_k(t)$ attains an absolute maximum at $t = a_2 - \frac{(n-k-1)}{(n-p)} (a_2-a_1)$ if $k = p \le n-1$ and at $t = a_1$ if $k+1 \le p \le n-1$. Further, for both functions the absolute maximum value is $E_{n,k}(a_2-a_1)^{n-k}$.

For $k = p+1, p+2, \ldots, n-1$ the pth derivative of any function $x(t)$, satisfying (3.20) will have $n-p-1$ zeros at $a_1$ and one zero at $a_2$. We define $x^{(p)}(t) = h(t)$ then, from Theorem 8.2 it follows that

$$|h^{(k)}(t)| \le \frac{k}{(n-p)(n-p-k)!} m(a_2-a_1)^{n-p-k}$$

$$= E_{n,p+k} \, m(a_2-a_1)^{n-p-k} .$$

For any function $x(t)$, satisfying (3.24) the result follows analogously.

In (8.11) the constants $E_{n,k}$ are the best possible as they are exact for the functions

$$x_1(t) = (t-a_1)^{n-1}[\frac{a_2-a_1}{n-p} - \frac{t-a_1}{n}] \quad \text{(for (3.20))}$$

$$x_2(t) = (a_2-t)^{n-1}[\frac{a_2-a_1}{n-p} - \frac{a_2-t}{n}] \quad \text{(for (3.24))}$$

and only for these functions upto a constant factor.

## COMMENTS AND BIBLIOGRAPHY

Theorem 8.1 is folklore in interpolation theory [8]. Besides being useful in answering questions connected with the error estimates in spline interpolation [11] these inequalities are also of immense value in the theory of BVPs which we shall take up in some of the following sections.

1. Agarwal, R. P. "Some inequalities for a function having n zeros", Proc. conf. General Inequalities 3, Ed. E. F. Beckenbach and W. Walter, ISNM 64, 371-378 (1983).
2. Agarwal, R. P. and Krishnamoorthy, P. R. "Boundary value problems for nth order ordinary differential equations", Bull. Inst. Math. Acad. Sinica 7, 211-230 (1979).
3. Agarwal, R. P. and Usmani, R. A. "Iterative methods for solving right focal point boundary value problems", J. Comp. Appl. Math. to appear.
4. Bessmertnyh, G. A. and Levin, A. Ju. "Some inequalities satisfied by differentiable functions of one variable", Soviet Math. Dokl. 3, 737-740 (1962).
5. Birkhoff, G. and Priver, A. "Hermite interpolation errors for derivatives", J. Math. Phys. 46, 440-447 (1967).
6. Brink, J. "Inequalities involving $\|f\|_p$ and $\|f^{(n)}\|_q$ for f with n zeros", Pacific J. Math. 42, 289-311 (1972).
7. Coppel, W. A. Disconjugacy, Lecture notes in Mathematics 220, Springer-Verlag, New York, 1971.
8. Davis, P. J. Interpolation and Approximation, Blaisdell Pub. Comp., Waltham, 1963.
9. Hukuhara, M. "On the zeros of solutions of linear ordinary differential equations", Sûgaku 15, 108-109 (1963).
10. Levin, A. Ju. "A bound for a function with monotonely distributed zeros of successive derivatives", Mat. Sb. 64 (106), 396-409 (1964).

88

11.  Schultz, M. H. <u>Spline Analysis</u>, Prentice-Hall, Inc. Englewood Cliffs, N. J., 1973.
12.  Tumura, M. "Kôkai Zyôbibunhôteisiki ni tuite", Kansû Hôteisiki <u>30</u>, 20-35 (1941).

# 9. EXISTENCE AND UNIQUENESS

The necessary and sufficient condition (2.15) for the existence of a unique solution of the BVP (2.1), (2.2) is only of theoretical value because it can be verified only for some trivial problems. Further, the method used in Theorem 7.2 to obtain explicit bounds for the existence of a solution of the BVP (7.1), (7.7), (7.8) leads to an insurmountable task for (2.1), (2.2). However, using the inequalities obtained in the previous section, it is possible to provide explicit bounds for the existence and uniqueness of the solutions even for the nonlinear nth order differential equation

$$(9.1) \qquad x^{(n)} = f(t,\underline{x})$$

together with each of the boundary conditions (2.3)-(2.9). In the differential equation (9.1), $\underline{x}$ stands for $(x, x', \ldots, x^{(q)})$, $0 \leq q \leq n-1$. Throughout, we shall assume that the function f is continuous at least in the interior of the domain of interest.

Theorem 9.1 [1]  Suppose that

(i)  $K_i > 0$, $0 \leq i \leq q$ are given real numbers and let Q be the maximum of $|f(t, x_0, x_1, \ldots, x_q)|$ on the compact set : $[a_1, a_r] \times D_0$, where

$$D_0 = \{(x_0, x_1, \ldots, x_q) : |x_i| \leq 2K_i, \ 0 \leq i \leq q\}$$

(ii)  $\max_{a_1 \leq t \leq a_r} |P_{n-1}^{(i)}(t)| \leq K_i$, $0 \leq i \leq q$

where $P_{n-1}(t)$ is the Hermite interpolating polynomial

$$P_{n-1}(t) = \sum_{i=1}^{r} \sum_{j=0}^{k_i} \sum_{\ell=0}^{k_i - j} \frac{1}{j! \ell!} \left[ \frac{(t-a_i)^{k_i+1}}{p(t)} \right]^{(\ell)}_{t=a_i} \times \frac{p(t)}{(t-a_i)^{k_i+1-j-\ell}} A_{j+1,i}$$

(iii)   $(a_r - a_1) \leq (K_i / Q C_{n,i})^{1/n-i}$, $0 \leq i \leq q$.

Then, the BVP (9.1), (2.4) has a solution in $D_0$.

Proof.   The set

$$B[a_1, a_r] = \{x(t) \in C^{(q)}[a_1, a_r] : \|x^{(i)}\| \leq 2K_i, 0 \leq i \leq q\}$$

where $\|x^{(i)}\| = \max\limits_{a_1 \leq t \leq a_r} |x^{(i)}(t)|$ is a closed convex subset of the Banach

space $C^{(q)}[a_1, a_r]$.   Consider an operator $T : C^{(q)}[a_1, a_r] \rightarrow C^{(n)}[a_1, a_r]$

as follows

(9.2)        $(Tx)(t) = P_{n-1}(t) + \displaystyle\int_{a_1}^{a_r} g_2(t,s) f(s, \underline{x}(s)) ds.$

Obviously, any fixed point of (9.2) is a solution of (9.1), (2.4).

We note that $(Tx)(t) - P_{n-1}(t)$ satisfies the conditions of Theorem 8.3, and

$$(Tx)^{(n)}(t) - P_{n-1}^{(n)}(t) = (Tx)^{(n)}(t) = f(t, \underline{x}(t)).$$

Thus, for all $x(t) \in B[a_1, a_r]$

$$\|(Tx)^{(n)}\| \leq Q$$

and

$$\|(Tx)^{(i)} - P_{n-1}^{(i)}\| \leq C_{n,i} Q (a_r - a_1)^{n-i}, \quad 0 \leq i \leq q$$

which also implies that

(9.3)        $\|(Tx)^{(i)}\| \leq \|P_{n-1}^{(i)}\| + C_{n,i} Q (a_r - a_1)^{n-i}$

$$\leq K_i + K_i = 2K_i, \quad 0 \leq i \leq q.$$

Thus, T maps $B[a_1,a_r]$ into itself. Further, the inequalities (9.3) imply that the sets $\{(Tx)^{(i)}(t) : x(t) \in B[a_1,a_r]\}$, $0 \leq i \leq q$ are uniformly bounded and equicontinuous on $[a_1,a_r]$. Hence, $\overline{TB[a_1,a_r]}$ is compact follows from the Ascoli-Arzela theorem. The Schauder fixed point theorem is applicable and a fixed point of (9.2) in $D_0$ exists.

**Theorem 9.2**   Suppose that the following hold

   (i)   condition (i) of Theorem 9.1 on $[a,b] \times D_0$

   (ii)   condition (ii) of Theorem 9.1, where $P_{n-1}(t)$ is the Abel-Gontscharoff interpolating polynomial

$$P_{n-1}(t) = \sum_{i=0}^{n-1} T_i(t)A_{i+1}$$

and

$$T_i(t) = \int_{a_1}^{t} dt_1 \int_{a_2}^{t_1} dt_2 \ldots \int_{a_i}^{t_{i-1}} dt_i, \quad 0 \leq i \leq n-1$$

   (iii)   $(b-a) \leq (\frac{K_i}{Q}(n-i)[\frac{n-i-1}{2}]![\frac{n-i}{2}]!)^{1/n-i}$, $0 \leq i \leq q$.

Then, the BVP (9.1), (2.5) has a solution in $D_0$.

**Theorem 9.3 [2]**   Suppose that the following hold

   (i)   condition (i) of Theorem 9.1 on $[a_1,a_2] \times D_0$

   (ii)   condition (ii) of Theorem 9.1, where $P_{n-1}(t)$ is the right focal point interpolating polynomial

$$P_{n-1}(t) = \sum_{i=0}^{k-1} \frac{(t-a_1)^i}{i!}A_i + \sum_{j=0}^{n-k-1}(\sum_{i=0}^{j}\frac{(t-a_1)^{k+i}(a_1-a_2)^{j-i}}{(k+i)!(j-i)!})A_{k+j}$$

   (iii)   $(a_2-a_1) \leq (K_i/QD_{n,i})^{1/n-i}$, $0 \leq i \leq q$.

Then, the BVP (9.1), (2.6) has a solution in $D_0$.

Theorem 9.4 [3]    Suppose that the following hold

  (i)   condition (i) of Theorem 9.3

  (ii)  condition (ii) of Theorem 9.1, where $P_{n-1}(t)$ is the $(n-1,p)$ interpolating polynomial

$$P_{n-1}(t) = \sum_{i=0}^{n-2} \frac{(t-a_1)^i}{1!} A_i + [B - \sum_{i=0}^{n-p-2} \frac{(a_2-a_1)^i}{i!} A_{p+i}] \frac{(n-p-1)!}{(n-1)!} \times$$

$$\frac{(t-a_1)^{n-1}}{(a_2-a_1)^{n-p-1}}$$

  (iii) $(a_2-a_1) \le (K_i/QE_{n,i})^{1/n-i}$, $0 \le i \le q$.

Then, the BVP (9.1), (2.7) has a solution in $D_0$.

Theorem 9.5 [3]    Suppose that the conditions of Theorem 9.4 are satisfied except that $P_{n-1}(t)$ is the $(p,n-1)$ interpolating polynomial

$$P_{n-1}(t) = \sum_{i=0}^{n-2} (-1)^i \frac{(a_2-t)^i}{i!} A_i$$

$$+ [B - \sum_{i=0}^{n-p-2} (-1)^{i+p} \frac{(a_2-a_1)^i}{i!} A_{p+i}] \frac{(n-p-1)!}{(n-1)!} \frac{(a_2-t)^{n-1}}{(a_2-a_1)^{n-p-1}}.$$

Then, the BVP (9.1), (2.8) has a solution in $D_0$.

    The proof of the above Theorems 9.2-9.5 is similar to that of Theorem 9.1.

    The BVP (9.1), (2.9) is a particular case of the following system

(9.4)                 $u'' = F(t,u,u')$

(9.5)                 $u(a_1) = \ell^1$,   $u(a_2) = \ell^2$

where $u(t)$ and $F(t,u,u')$ are $m \times 1$ vectors with components $u_i(t)$ and $f_i(t,u,u')$, $1 \leq i \leq m$ respectively.

The following result for any convenient norm $\|u\|$ of $u \in R^m$ can be deduced from Theorem 7.2.

**Theorem 9.6**  Suppose that

  (i)  $K_0 > 0$, $K_1 > 0$ are given real numbers and $Q$ is the maximum

  of $\|F(t,u^0,u^1)\|$ on the compact set : $[a_1,a_2] \times D_0$, where

  $D_0 = \{(u^0,u^1) : \|u^0\| \leq 2K_0, \|u^1\| \leq 2K_1\}$

  (ii)  $\|\ell^1\| \leq K_0$, $\|\ell^2\| \leq K_0$, $\|\ell^1 - \ell^2\|/(a_2-a_1) \leq K_1$

  (iii)  $(a_2-a_1) = \min\left\{(\dfrac{8K_0}{Q})^{\frac{1}{2}}, \dfrac{2K_1}{Q}\right\}$.

Then, the BVP (9.4), (9.5) has a solution in $D_0$.

Hereafter, we shall prove results only for the BVP (9.1), (2.4) whereas, for the other problems analogous results can easily be stated.

**Corollary 9.7**  Suppose that the conditions of Theorem 9.1 are satisfied. Then, for any $\varepsilon > 0$ there is a solution $x(t)$ of the BVP (9.1), (2.4) such that $|x^{(i)}(t) - P_{n-1}^{(i)}(t)| < \varepsilon$, $0 \leq i \leq n-1$ provided $(a_r-a_1)$ is sufficiently small.

**Corollary 9.8**  Suppose that the conditions (i), (iii) of Theorem 9.1 are satisfied. Then, for any $g(t) \in C^{(n-1)}[a_1,a_r]$ the differential equation (9.1) together with

$$x^{(j)}(a_i) = g^{(j)}(a_i); \quad 1 \leq i \leq r, \ 0 \leq j \leq k_i$$

has a solution if

$$\sum_{j=i}^{n-1} M_j (a_r - a_1)^{j-i} \le K_i, \quad 0 \le i \le q$$

where

$$M_j = \max_{a_1 \le t \le a_r} |g^{(j)}(t)|, \quad 0 \le j \le n-1.$$

Proof. We need to verify that the condition (ii) of Theorem 9.1 is satisfied. For this, in $P_{n-1}(t)$ we take $A_{j+1,i} = g^{(j)}(a_i)$; $1 \le i \le r$, $0 \le j \le k_i$. Then, the function $\phi(t) = g(t) - P_{n-1}(t)$ has n zeros in $[a_1, a_r]$. Thus, from the generalized Rolle's theorem $\phi^{(k)}(t)$, $1 \le k \le n-1$ vanishes at least n-k times in $(a_1, a_r)$. Let $t_k \in (a_1, a_r)$ be any zero of $\phi^{(k)}(t)$, then

$$|P_{n-1}^{(n-1)}(t)| = |P_{n-1}^{(n-1)}(t_{n-1})| = |g^{(n-1)}(t_{n-1})| \le \max_{a_1 \le t \le a_r} |g^{(n-1)}(t)| = M_{n-1}$$

and

$$|P_{n-1}^{(n-2)}(t)| \le |P_{n-1}^{(n-2)}(t_{n-2})| + \left| \int_{t_{n-2}}^{t} |P_{n-1}^{(n-1)}(s)| ds \right|$$

$$= |g^{(n-2)}(t_{n-2})| + \left| \int_{t_{n-2}}^{t} |g^{(n-1)}(t_{n-1})| ds \right|$$

$$\le M_{n-2} + M_{n-1}(a_r - a_1).$$

Using the same argument repeatedly, we obtain

$$|P_{n-1}^{(i)}(t)| \le \sum_{j=i}^{n-1} M_j (a_r - a_1)^{j-i} .$$

Corollary 9.9 Suppose that the function $f(t, x_0, x_1, \ldots, x_q)$ is such that

(9.6) $$|f(t, x_0, x_1, \ldots, x_q)| \le L + \sum_{i=0}^{q} L_i |x_i|^{\alpha(i)}$$

for all $(t,x_0,x_1,\ldots,x_q) \in [a_1,a_r] \times R^{q+1}$, where $0 \leq \alpha(i) < 1$, $L$ and $L_i$, $0 \leq i \leq q$ are nonnegative constants. Then, the BVP (9.1), (2.4) has a solution.

**Proof.** Condition (9.6) implies that on $[a_1,a_r] \times D_0$

$$|f(t,x_0,x_1,\ldots,x_q)| \leq L + \sum_{i=0}^{q} L_i (2K_i)^{\alpha(i)} = Q_1 \text{ (say)}.$$

Now, Theorem 9.1 is applicable by choosing $K_i$, $0 \leq i \leq q$ so large that condition (ii) holds and

$$C_{n,i} Q_1 (a_r - a_1)^{n-i} \leq K_i, \quad 0 \leq i \leq q.$$

Theorem 9.1 is a local existence theorem whereas Corollary 9.9 does not require any condition on the length of the interval or the boundary conditions. The question : what happens if $\alpha(i) = 1$, $0 \leq i \leq q$ in (9.6) is considered in the next result.

**Theorem 9.10 [1]** Suppose that the function $f(t,x_0,x_1,\ldots,x_q)$ is such that

(9.7)
$$|f(t,x_0,x_1,\ldots,x_q)| \leq L + \sum_{i=0}^{q} L_i |x_i|$$

for all $(t,x_0,x_1,\ldots,x_q) \in [a_1,a_r] \times D_1$, where

$$D_1 = \{(x_0,x_1,\ldots,x_q) : |x_i| \leq \max_{a_1 \leq t \leq a_r} |P_{n-1}^{(i)}(t)| + C_{n,i} (a_r-a_1)^{n-i} \frac{L+c}{1-\theta}, \ 0 \leq i \leq q\}$$

and

$$c = \max_{a_1 \leq t \leq a_r} \sum_{i=0}^{q} L_i |P_{n-1}^{(i)}(t)|$$

(9.8)
$$\theta = \sum_{i=0}^{q} C_{n,i} L_i (a_r-a_1)^{n-i} < 1.$$

Then, the BVP (9.1), (2.4) has a solution in $D_1$.

Proof.   The BVP (9.1), (2.4) is equivalent to the problem

$$(9.9) \qquad y^{(n)}(t) = f(t, \underline{y}(t) + \underline{P}_{n-1}(t))$$

$$(9.10) \qquad y^{(j)}(a_i) = 0; \ 1 \le i \le r, \ 0 \le j \le k_i.$$

We define M as the set of functions n times continuously differentiable on $[a_1, a_r]$ satisfying the boundary conditions (9.10).  If we introduce in M the norm $\|y\| = \max\limits_{a_1 \le t \le a_r} |y^{(n)}(t)|$, then M becomes a

Banach space.  We shall show that the mapping $T : M \to M$ defined by

$$(9.11) \qquad (Ty)(t) = \int_{a_1}^{a_r} g_2(t,s) f(s, \underline{y}(s) + \underline{P}_{n-1}(s)) ds$$

maps the ball $S = \{y(t) \ e \ M : \|y\| \le \frac{L+c}{1-\theta}\}$ into itself.  For this, let $y(t) \ e \ S$ then from Theorem 8.3, we have

$$|y^{(i)}(t)| \le C_{n,i}(a_r - a_1)^{n-i} \frac{L+c}{1-\theta} , \ 0 \le i \le q$$

and hence

$$|y^{(i)}(t) + P_{n-1}^{(i)}(t)| \le \max_{a_1 \le t \le a_r} |P_{n-1}^{(i)}(t)| + C_{n,i}(a_r - a_1)^{n-i} \frac{L+c}{1-\theta}, \ 0 \le i \le q$$

which implies that $(t, \underline{y}(t) + \underline{P}_{n-1}(t)) \ e \ [a_1, a_r] \times D_1$.

Further, from (9.11) we have

$$\| (Ty) \| = \max_{a_1 \le t \le a_r} |f(t, \underline{y}(t) + \underline{P}_{n-1}(t))|$$

$$\leq L + \sum_{i=0}^{q} L_i |y^{(i)}(t) + P_{n-1}^{(i)}(t)|$$

$$\leq L + c + \sum_{i=0}^{q} L_i C_{n,i} (a_r - a_1)^{n-i} \frac{L+c}{1-\theta}$$

$$= L + c + \theta \frac{L+c}{1-\theta}$$

$$= \frac{L+c}{1-\theta} .$$

Thus, it follows from Schauder's fixed point theorem that T has a fixed point in S. This fixed point $y(t)$ is a solution of (9.9), (9.10) and hence the BVP (9.1), (2.4) has a solution $x(t) = y(t) + P_{n-1}(t)$.

**Theorem 9.11 [4]**   Suppose that the BVP (9.1), (3.8) has a nontrivial solution $x(t)$ and the condition (9.7) with $L = 0$ is satisfied for all $(t, x_0, x_1, \ldots, x_q) \in [a_1, a_r] \times D_2$, where

$$D_2 = \{(x_0, x_1, \ldots, x_q) : |x_i| \leq C_{n,i} (a_r - a_1)^{n-i} m, \ 0 \leq i \leq q\}$$

and $m = \max\limits_{a_1 \leq t \leq a_r} |x^{(n)}(t)|$. Then, it is necessary that $\theta \geq 1$.

**Proof.**  Since $x(t)$ is a nontrivial solution of (9.1), (3.8) it is necessary that $m \neq 0$, and Theorem 8.3 implies that $(t, \underline{x}(t)) \in [a_1, a_r] \times D_2$. Thus, we have

$$m = \max_{a_1 \leq t \leq a_2} |x^{(n)}(t)| = \max_{a_1 \leq t \leq a_r} |f(t, \underline{x}(t))|$$

$$\leq \max_{a_1 \leq t \leq a_r} \sum_{i=0}^{q} L_i |x^{(i)}(t)|$$

$$\leq \sum_{i=0}^{q} L_i C_{n,i} (a_r - a_1)^{n-i} m$$

$$= \theta m$$

and hence $\theta \geq 1$.

Conditions of Theorem 9.11 ensure that in (9.7) at least one of the $L_i$, $0 \leq i \leq q$ will not be zero, otherwise on $[a_1, a_r]$ the solution $\dot{x}(t)$ will coincide with a polynomial of degree at most $(n-1)$ and will not be a nontrivial solution of (9.1), (3.8). Further, $x(t) \equiv 0$ is obviously a solution of (9.1), (3.8) and, if $\theta < 1$, then it is also unique.

Theorem 9.12 [1]  Suppose that for all $(t, x_0, x_1, \ldots, x_q), (t, y_0, y_1, \ldots, y_q)$ $\epsilon$ $[a_1, a_r] \times D_1$ the function f satisfies the Lipschitz condition

$$(9.12) \qquad |f(t, x_0, x_1, \ldots, x_q) - f(t, y_0, y_1, \ldots, y_q)| \leq \sum_{i=0}^{q} L_i |x_i - y_i|$$

where $L = \max_{a_1 \leq t \leq a_r} |f(t, 0, 0, \ldots, 0)|$. Then, the BVP (9.1), (2.4) has a unique solution in $D_1$.

Proof.  Lipschitz condition (9.12) implies (9.7) and hence the existence of a solution in $D_1$ follows from Theorem 9.10.  To show the uniqueness let $y_1(t)$ and $y_2(t)$ be two solutions of (9.9), (9.10) in $D_1$.  Then, since $(t, \underline{y}_1(t) + \underline{P}_{n-1}(t))$, $(t, \underline{y}_2(t) + \underline{P}_{n-1}(t))$ $\epsilon$ $[a_1, a_r] \times D_1$ from (9.9), (9.12) and Theorem 8.3, we get

$$|y_1^{(n)}(t) - y_2^{(n)}(t)| \leq \max_{a_1 \leq t \leq a_r} \sum_{i=0}^{q} L_i |y_1^{(i)}(t) - y_2^{(i)}(t)|$$

$$\leq \theta \max_{a_1 \leq t \leq a_r} |y_1^{(n)}(t) - y_2^{(n)}(t)|.$$

Since $\theta < 1$, we find that $y_1^{(n)}(t) \equiv y_2^{(n)}(t)$ for all $t \epsilon [a_1, a_r]$.  Thus, $y_1(t) \equiv y_2(t)$ follows from the boundary conditions (9.10).

Lemma 9.13 [6]  Suppose $x(t) \in C^{(n-1)}[a,b]$ has at least n zeros in [a,b]. Then, we can find points $t_0, t_1, \ldots, t_{2n-2}$ such that

$$a \le t_0 \le t_1 \le \cdots \le t_{2n-2} \le b$$

and

$$0 = x(t_0) = x'(t_1) = \ldots = x^{(n-2)}(t_{n-2}) = x^{(n-1)}(t_{n-1}) =$$

$$x^{(n-2)}(t_n) = \ldots = x'(t_{2n-3}) = x(t_{2n-2}).$$

Proof.  Let $t_1^0, t_2^0, \ldots, t_n^0$ be n zeros of $x(t)$ such that $a \le t_1^0 \le t_2^0 \le \ldots \le t_n^0 \le b$. By Rolle's theorem we can find n-1 zeros $t_1^1, t_2^1, \ldots, t_{n-1}^1$ of $x'(t)$ such that $t_i^0 \le t_i^1 \le t_{i+1}^0$, $1 \le i \le n-1$. Repeating this process we obtain finally a zero $t_1^{n-1}$ of $x^{(n-1)}(t)$ between two zeros $t_1^{n-2}$, $t_2^{n-2}$ of $x^{(n-2)}(t)$. The points

$$t_1^0, t_1^1, \ldots, t_1^{n-2}, t_1^{n-1}, t_2^{n-2}, \ldots, t_{n-1}^1, t_n^0$$

satisfy the requirements of the lemma.

Theorem 9.14  Suppose that the function f satisfies the Lipschitz condition (9.12) on $[a,b] \times R^{q+1}$. Further, we assume that

$$(9.13) \qquad \beta = \sum_{k=n-q}^{n} \frac{1}{2^k k [\frac{k-1}{2}]! [\frac{k}{2}]!} L_{n-k}(b-a)^k \le 1.$$

Then, the problem (9.1), (2.4) has a unique solution.

Proof.  We shall prove only the uniqueness part. Suppose on the contrary that the problem (9.1), (2.4) has two solutions $x(t)$ and $y(t)$. Then, the function $h(t) = x(t) - y(t)$ satisfies (3.8). Hence, by Lemma 9.13 with $t_{n-1} = c$, $h(t)$ satisfies (8.6) on a subinterval $[a,c]$, and same conditions with the inequalities satisfied by $t_0, t_1, \ldots, t_{n-1}$ reversed on the complementary subinterval $[c,b]$. One of these two subintervals, say, $[a,c]$ has length at most $(b-a)/2$. Moreover the interval $[a,c]$ is nondegenerate, since $h(t)$ cannot have a zero of multiplicity $n$. Applying Theorem 8.4 to this interval, we obtain

$$(9.14) \qquad \left|h^{(n-k)}(t)\right| \le \mu \, \frac{(b-a)^k}{2^k k[\frac{k-1}{2}]![\frac{k}{2}]!}$$

where $\mu = \max\limits_{a \le t \le c} \left|h^{(n)}(t)\right|$. But, for some $\tau \in [a,c]$

$$\mu = \left|h^{(n)}(\tau)\right| = \left|f(\tau,\underline{x}(\tau)) - f(\tau,\underline{y}(\tau))\right|$$

$$\le \sum_{k=n-q}^{n} L_{n-k}\left|h^{(n-k)}(\tau)\right|$$

$$\le \mu\beta.$$

Evidently $\mu > 0$, since otherwise $h(t)$ would coincide on $[a,c]$ with a polynomial of degree $m < n$ and $h^{(m)}(t)$ would not vanish on $[a,c]$. Hence, $\beta \ge 1$. It only remains to exclude the possibility of equality. At least one of the numbers $L_0, L_1, \ldots, L_q$ is different from zero, since otherwise $h(t)$ would be a polynomial of degree less than $n$ and could not have $n$ zeros. Thus, if $\beta = 1$ then equality must hold in (9.14) for at least one value of $k$. This is possible only if $h(t)$ coincides on $[a,c]$ with a polynomial of degree $n$. But, we can take $\tau$ to be any point of $[a,c]$, and $\left|h^{(n-k)}(\tau)\right|$ is not a constant on $[a,c]$ for any $k = n-q$, $n-q+1,\ldots,n$. Therefore, in this case also, we have $\beta > 1$.

**Theorem 9.15** [7]  The BVP (2.1), (2.4) has a unique solution provided $\max\{Q_1, Q_2\} \le 1$, where

$$Q_1 = (\exp \int_a^{(a+b)/2} |p_1(s)|ds) \sum_{k=1}^{n-1} \frac{(b-a)^k \int_a^{(a+b)/2} |p_{k+1}(s)|ds}{2^k k[\frac{k-1}{2}]![\frac{k}{2}]!}$$

and

$$Q_2 = (\exp \int_{(a+b)/2}^{b} |p_1(s)| ds) \sum_{k=1}^{n-1} \frac{(b-a)^k \int_{(a+b)/2}^{b} |p_{k+1}(s)| ds}{2^k k [\frac{k-1}{2}]! [\frac{k}{2}]!} .$$

**Proof.** From the considerations of Section 2, it suffices to show that the BVP (2.1), (2.4) has at most one solution. For this, we define the function h(t) and the subintervals [a,c], [c,b] as in the proof of Theorem 9.14. The resulting differential equation for h(t) can be written as

$$(h^{(n-1)}(t)\exp(\int_a^t p_1(s)ds))' = -\sum_{k=1}^{n-1} p_{k+1}(t)h^{(n-k-1)}(t)\exp(\int_a^t p_1(s)ds).$$

Let $M = \max_{a \le t \le c} |h^{(n-1)}(t)|$, and $|h^{(n-1)}(r)| = M$, $r \in [a,c]$. An integration over the interval [r,c] gives

$$M \exp(\int_a^r p_1(s)ds) < \sum_{k=1}^{n-1} \max_{a \le t \le c} |h^{(n-k-1)}(t)| \int_r^c (\exp \int_a^t p_1(s)ds) |p_{k+1}(t)| dt.$$

By Theorem 8.4, and

$$\exp(\int_r^t p_1(s)ds) \le \exp(\int_a^c |p_1(s)| ds), \quad a \le r \le t \le c$$

we get

(9.15)    $$1 < (\exp \int_a^c |p_1(s)| ds) \sum_{k=1}^{n-1} \frac{(c-a)^k}{k [\frac{k-1}{2}]! [\frac{k}{2}]!} \int_a^c |p_{k+1}(t)| dt.$$

Similarly, we find

(9.16)    $$1 < (\exp \int_c^b |p_1(s)| ds) \sum_{k=1}^{n-1} \frac{(b-c)^k}{k [\frac{k-1}{2}]! [\frac{k}{2}]!} \int_c^b |p_{k+1}(t)| dt.$$

Since, either $c \leq (a+b)/2$ or $c \geq (a+b)/2$, i.e., $(c-a) \leq (b-a)/2$ or $(b-c) \leq (b-a)/2$, (9.15) and (9.16) imply that $Q_1 > 1$ or $Q_2 > 1$.

Example 9.1 [5]  For the BVP

$$(9.17) \qquad x^{(4)} = x^2 \sin x + \sin t$$

$$(9.18) \qquad x(0) = 1, \ x'(0) = x(1) = x'(1) = 0,$$

we have $P_3(t) = (1-t)^2(1+2t)$, $Q = 4K_0^2 + 1$.  Thus, the conditions of Theorem 9.1 are satisfied provided

$$1 \leq K_0 \qquad \text{and} \qquad 1 < \frac{384 \, K_0}{4K_0^2 + 1} \ .$$

Hence, there exists at least one solution of (9.17), (9.18) in the region

$$S = \{(t,x) : 0 \leq t \leq 1, \ |x| \leq 2K_0 \text{ where } 1 \leq K_0 \leq 95.99739576...\}.$$

Example 9.2 [5]  For the BVP

$$x^{(4)} = |x|^{\frac{1}{2}}\sin e^x + e^{-t^2}$$

$$x(0) = 1, \ x'(0) = 2, \ x(1) = 1, \ x'(1) = 0$$

Corollary 9.9 ensures the existence of at least one solution in the region $S = \{(t,x) : 0 \leq t \leq 1, \ |x| < \infty\}$.

Example 9.3 [5]  Consider the differential equation

$$(9.19) \qquad x^{(4)} = x \sin x + e^{-t^2}$$

together with the boundary conditions (9.18).  For the function

$x \sin x + e^{-t^2}$ condition (9.7) is satisfied with $L = 1$, $L_0 = 1$ and (9.8) is obvious. Thus, the BVP (9.19), (9.18) has at least one solution $x(t)$ in the region

$$S = \{(t,x) : 0 \le t \le 1, \; |x| \le 1.005222\}.$$

**Example 9.4** [5]  Consider the differential equation

$$x^{(4)} = x \sin x^2$$

together with $x(a_1) = x'(a_1) = x(a_2) = x'(a_2) = 0$. For this BVP, Theorem 9.11 ensures that $x(t) \equiv 0$ is the only solution as long as $(a_2 - a_1) < 4.4267$.

## COMMENTS AND BIBLIOGRAPHY

Results of this section provide an easier test for the local existence and uniqueness of the solutions. In conclusion, if f is continuous and $\det(\ell_i[t^{j-1}]) \ne 0$, $1 \le i$, $j \le n$ then there exists a sufficiently small interval (b-a) on which the BVP (9.1), (2.2) has a solution. Further, if f satisfies the Lipschitz condition (9.12), then this solution is unique as well. Theorem 9.14 is due to Opial [9], but its proof is modelled after Levin [8] and Coppel [6]. Theorem 9.15 corrects an error in the result given in [10].

1. Agarwal, R. P. "Boundary value problems for higher order integro-differential equations",Nonlinear Analysis : Theory, Methods and Appl. 7, 259-270 (1983).
2. Agarwal, R. P. and Usmani, R. A. "Iterative methods for solving right focal point boundary value problems", J. Comp. Appl. Math. to appear.
3. Agarwal, R. P. "Boundary value problems for higher order differential equations", Bull. Inst. Math. Acad. Sinica 9, 47-61 (1981).
4. Agarwal, R. P. "Necessary conditions for the existence of solutions of multi-point boundary value problems", Bull. Inst. Math. Acad. Sinica 12, 11-16 (1984).

5.  Agarwal, R. P. and Chow, Y. M. "Iterative methods for a fourth order boundary value problem", J. Comp. Appl. Math. 10, 203-217 (1984).

6.  Coppel, W. A. Disconjugacy, Lecture notes in Mathematics 220, Springer-Verlag, New York, 1971.

7.  Hartman, P. "On disconjugacy criteria", Proc. Amer. Math. Soc. 24, 374-381 (1970).

8.  Levin, A. Ju. "A bound for a function with monotonely distributed zeros of successive derivatives", Mat. Sb. 64 (106), 396-409 (1964).

9.  Opial, Z. "Linear problems for systems of nonlinear differential equations", J. Diff. Equs. 3, 580-594 (1967).

10. Zaĭceva, G. S. "A multipoint boundary value problem", Soviet Math. Dokl. 8, 1183-1185 (1967).

## 10. PICARD'S AND APPROXIMATE PICARD'S METHOD

The Picard method of successive approximations mentioned in Section 7 for the BVP (7.1), (7.2) has an important characteristic, that it is constructive; moreover, bounds of the difference between iterates and the solution are easily available. In this section we shall discuss this method only for the BVP (9.1), (2.4). For other BVPs analogous results can be stated without much difficulty. For this, we need

**Theorem 10.1** (Contraction mapping principle) [3]    Let B be a Banach space, and let $r > 0$; $S(x_0, r) = \{x \in B : \|x - x_0\| \le r\}$. Let T map $\bar{S}(x_0, r)$ into B and

(i)    for all $x, y \in \bar{S}(x_0, r)$, $\|Tx - Ty\| \le \alpha \|x-y\|$ where $0 \le \alpha < 1$

(ii)    $r_0 = (1-\alpha)^{-1} \|Tx_0 - x_0\| \le r$.

Then, the following hold

(1)    T has a fixed point $x^*$ in $\bar{S}(x_0, r_0)$

(2)    $x^*$ is the unique fixed point of T in $\bar{S}(x_0, r)$

(3)    the sequence $\{x_m\}$ defined by

$$(10.1) \qquad x_{m+1} = Tx_m \,; \quad m = 0, 1, \ldots$$

converges to $x^*$ with

$$\|x^* - x_m\| \le \alpha^m r_0$$

(4)    for any $x \in \bar{S}(x_0, r_0)$, $x^* = \lim_{m \to \infty} T^m x$

(5)    any sequence $\{\bar{x}_m\}$ such that $\bar{x}_m \in \bar{S}(x_m, \alpha^m r_0)$; $m = 0, 1, \ldots$ converges to $x^*$.

**Remark 10.1**    If we define $\bar{S}(x_0, \infty) = B$, then Theorem 10.1 guarantees the existence of $x^*$ in $\bar{S}(x_0, r_0)$ and its uniqueness in B.

**Remark 10.2**  In Theorem 10.1, condition (ii) forces T to map $\bar{S}(x_0, r_0)$ into itself. Indeed, if $x \in \bar{S}(x_0, r_0)$, then $\|x - x_0\| \leq r_0$ and from the triangle inequality, we have

$$\|Tx - x_0\| \leq \|Tx - Tx_0\| + \|Tx_0 - x_0\|$$

$$\leq \alpha \|x - x_0\| + (1-\alpha) r_0$$

$$\leq \alpha r_0 + (1-\alpha) r_0 = r_0.$$

However, $T\bar{S}(x_0, r) \subseteq \bar{S}(x_0, r)$ need not imply (ii). For example, let $B = R$, $\|x\| = |x|$, $Tx = \frac{2}{3} + \frac{1}{3} x^2$, $x_0 = 0$, $r = 1$. Obviously, $T\bar{S}(0,1) = T[-1,1] \subseteq [-1,1]$, and since $x_1 = \frac{2}{3}$, $\alpha = \frac{2}{3}$, we get $r_0 = 2 > r$.

**Theorem 10.2**  In Theorem 10.1, if $T\bar{S}(x_0, r) \subseteq \bar{S}(x_0, r)$, then condition (ii) can be omitted. However, then

(1)   T has a unique fixed point $x^*$ in $\bar{S}(x_0, r)$

(2)   the sequence $\{x_m\}$ defined by (10.1) converges to $x^*$ with

$$\|x^* - x_m\| \leq \alpha^m r$$

(3)   for any $x \in \bar{S}(x_0, r)$, $x^* = \lim_{m \to \infty} T^m x$

(4)   any sequence $\{\bar{x}_m\}$ such that $\bar{x}_m \in \bar{S}(x_m, \alpha^m r)$; $m = 0, 1, \ldots$ converges to $x^*$.

**Definition 10.1**  A function $\bar{x}(t) \in C^{(n)}[a_1, a_r]$ is called an approximate solution of (9.1), (2.4) if there exist $\delta$ and $\varepsilon$ nonnegative constants such that

$$(10.2) \qquad \max_{a_1 \leq t \leq a_r} |\bar{x}^{(n)}(t) - f(t, \bar{x}(t))| \leq \delta$$

and

$$(10.3) \qquad \max_{a_1 \leq t \leq a_r} |P_{n-1}^{(i)}(t) - \bar{P}_{n-1}^{(i)}(t)| \leq \varepsilon C_{n,i} (a_r - a_1)^{n-i}, \quad 0 \leq i \leq q$$

where $P_{n-1}(t)$ and $\bar{P}_{n-1}(t)$ are the polynomials of degree $(n-1)$ satisfying (2.4) and

(10.4) $$\bar{P}_{n-1}^{(j)}(a_i) = \bar{x}^{(j)}(a_i); \; 1 \leq i \leq r, \; 0 \leq j \leq k_i$$

respectively.

Inequality (10.2) means that there exists a continuous function $\eta(t)$ such that

(10.5) $$\bar{x}^{(n)}(t) = f(t,\bar{x}(t)) + \eta(t)$$

where $\displaystyle\max_{a_1 \leq t \leq a_r} |\eta(t)| \leq \delta$.

Thus, the approximate solution $\bar{x}(t)$ can be expressed as

(10.6) $$\bar{x}(t) = \bar{P}_{n-1}(t) + \int_{a_1}^{a_r} g_2(t,s)[f(s,\bar{x}(s)) + \eta(s)]ds.$$

In what follows, we shall consider the Banach space $B = C^{(q)}[a_1,a_r]$ and for all $x(t) \in B$

(10.7) $$\|x\| = \max_{0 \leq j \leq q} \left\{ \frac{C_{n,0}(a_r-a_1)^j}{C_{n,j}} \max_{a_1 \leq t \leq a_r} |x^{(j)}(t)| \right\}.$$

Theorem 10.3 [1]  With respect to (9.1), (2.4) we assume that there exists an approximate solution $\bar{x}(t)$ and

    (i)  the function f satisfies the Lipschitz condition (9.12) on $[a_1,a_r] \times D_3$, where

$$D_3 = \{(x_0,x_1,\ldots,x_q) : |x_j - \bar{x}^{(j)}(t)| \leq N \frac{C_{n,j}}{C_{n,0}(a_r-a_1)^j}, \; 0 \leq i \leq q\}$$

    (ii)  $\theta < 1$

(10.8)  (iii) $(1-\theta)^{-1}(\varepsilon+\delta)C_{n,0}(a_r-a_1)^n \leq N$.

Then, the following hold

    (1)    there exists a solution $x^*(t)$ of (9.1), (2.4) in $\bar{S}(\bar{x}, N_0)$

    (2)    $x^*(t)$ is the unique solution of (9.1), (2.4) in $\bar{S}(\bar{x}, N)$

    (3)    the Picard sequence $\{x_m(t)\}$ defined by

$$(10.9) \qquad x_{m+1}(t) = P_{n-1}(t) + \int_{a_1}^{a_r} g_2(t,s) f(s, \underline{x}_m(s)) ds$$

$$x_0(t) = \bar{x}(t); \; m = 0,1,\ldots$$

converges to $x^*(t)$ with

$$\|x^* - x_m\| \le \theta^m N_0$$

    (4)    for $x_0(t) = x(t) \; \epsilon \; \bar{S}(\bar{x}, N_0)$ the iterative process (10.9) converges to $x^*(t)$

    (5)    any sequence $\{\bar{x}_m(t)\}$ such that $\bar{x}_m(t) \; \epsilon \; \bar{S}(x_m, \theta^m N_0)$; $m = 0,1,\ldots$ converges to $x^*(t)$

where $N_0 = (1-\theta)^{-1} \|x_1 - \bar{x}\|$.

**Proof.** We shall show that the operator $T : \bar{S}(\bar{x}, N) \to C^{(n)}[a_1, a_r]$ defined in (9.2) satisfies the conditions of Theorem 10.1. Let $x(t) \; \epsilon \; \bar{S}(\bar{x}, N)$, then from the definition of norm (10.7), we have

$$\|x - \bar{x}\| = \max_{0 \le j \le q} \left\{ \frac{C_{n,0}(a_r - a_1)^j}{C_{n,j}} \max_{a_1 \le t \le a_r} |x^{(j)}(t) - \bar{x}^{(j)}(t)| \right\} \le N$$

which implies that

$$|x^{(j)}(t) - \bar{x}^{(j)}(t)| \le N \frac{C_{n,j}}{C_{n,0}(a_r - a_1)^j} , \quad 0 \le j \le q$$

and hence $x(t) \; \epsilon \; D_3$. Further, if $x(t)$, $y(t) \; \epsilon \; \bar{S}(\bar{x}, N)$, then $((Tx)(t) - (Ty)(t))$ satisfies the conditions of Theorem 8.3, and we get

$$|(Tx)^{(j)}(t)-(Ty)^{(j)}(t)| \leq C_{n,j}(a_r-a_1)^{n-j} \times \max_{a_1 \leq t < a_r} |f(t,\underline{x}(t))-f(t,\underline{y}(t))|$$

$$\leq C_{n,j}(a_r-a_1)^{n-j} \times \max_{a_1 \leq t < a_r} \sum_{i=0}^{q} L_i |x^{(i)}(t)-y^{(i)}(t)|$$

$$\leq C_{n,j}(a_r-a_1)^{n-j} \sum_{i=0}^{q} L_i \frac{C_{n,i}}{C_{n,0}(a_r-a_1)^i} \|x-y\|$$

and hence

$$\frac{C_{n,0}(a_r-a_1)^j}{C_{n,j}} |(Tx)^{(j)}(t) - (Ty)^{(j)}(t)|$$

$$\leq \sum_{i=0}^{q} L_i C_{n,i}(a_r-a_1)^{n-i} \|x-y\| , \quad 0 \leq j \leq q$$

from which it follows that

$$\|Tx - Ty\| \leq \theta \|x-y\| .$$

Next, from (9.2) and (10.6), we have

$$(T\bar{x})(t) - \bar{x}(t) = (Tx_0)(t) - x_0(t)$$

$$(10.10) \qquad = P_{n-1}(t) - \bar{P}_{n-1}(t) - \int_{a_1}^{a_r} g_2(t,s)\eta(s)ds.$$

The function $z(t) = -\int_{a_1}^{a_r} g_2(t,s)\eta(s)ds$ satisfies the conditions of

Theorem 8.3, and $z^{(n)}(t) = -\eta(t)$, thus

$$\max_{a_1 \leq t < a_r} |z^{(n)}(t)| = \max_{a_1 \leq t < a_r} |\eta(t)| \leq \delta$$

and hence

$$|z^{(j)}(t)| \leq C_{n,j}(a_r-a_1)^{n-j}\delta, \quad 0 \leq j \leq q.$$

Using these inequalities and (10.3) in (10.10), we obtain

$$\left| (Tx_0)^{(j)}(t) - x_0^{(j)}(t) \right| \leq (\varepsilon+\delta)C_{n,j}(a_r-a_1)^{n-j}$$

which is same as

$$\frac{C_{n,0}(a_r-a_1)^j}{C_{n,j}} \left| (Tx_0)^{(j)}(t) - x_0^{(j)}(t) \right| \leq (\varepsilon+\delta)C_{n,0}(a_r-a_1)^n$$

and hence

$$(10.11) \qquad \| Tx_0 - x_0 \| \leq (\varepsilon+\delta)C_{n,0}(a_r-a_1)^n.$$

Thus, from (10.8) it follows that

$$(1-\theta)^{-1} \| Tx_0 - x_0 \| \leq N.$$

Hence, the conditions of Theorem 10.1 are satisfied and conclusions (1)-(5) follow.

**Theorem 10.4**  Let the conditions of Theorem 10.3 be satisfied except that $\bar{x}(t) = P_{n-1}(t)$ and (10.8) is replaced by

$$(10.12) \qquad C_{n,0}M(a_r-a_1)^n \leq N$$

where $M \geq |f(t,\underline{x})|$ for all $(t,x) \in [a_1,a_r] \times \bar{S}(P_{n-1},N)$.

Then, the following hold

(1)   there exists a unique solution $x^*(t)$ of (9.1), (2.4) in $\bar{S}(P_{n-1},N)$

(2)   the sequence $\{x_m(t)\}$ defined by (10.9) with $x_0(t) = P_{n-1}(t)$ converges to $x^*(t)$ and

$$\| x^* - x_m \| \leq \theta^m N$$

(3)    for $x_0(t) = x(t) \; \epsilon \; \bar{S}(P_{n-1},N)$ the iterative process (10.9) converges to $x^*(t)$

(4)    any sequence $\{\bar{x}_m(t)\}$ such that $\bar{x}_m(t) \; \epsilon \; \bar{S}(x_m, \theta^m N)$; $m = 0,1,\ldots$ converges to $x^*(t)$.

**Proof.** Following the proof of Theorem 10.3, for any $x(t) \; \epsilon \; \bar{S}(P_{n-1},N)$, we have

$$|(Tx)^{(j)}(t) - P_{n-1}^{(j)}(t)| \leq C_{n,j}(a_r - a_1)^{n-j} \times \max_{a_1 \leq t \leq a_r} |f(t,\underline{x}(t))|$$

$$\leq M C_{n,j}(a_r - a_1)^{n-j}$$

or

$$\frac{C_{n,0}(a_r - a_1)^j}{C_{n,j}} |(Tx)^{(j)}(t) - P_{n-1}^{(j)}(t)| \leq C_{n,0} M(a_r - a_1)^n, \quad 0 \leq j \leq q$$

which implies that

$$\|Tx - P_{n-1}\| \leq C_{n,0} M(a_r - a_1)^n.$$

Thus, from (10.12) we find that $(Tx)(t) \epsilon \bar{S}(P_{n-1},N)$. Hence, the conditions of Theorem 10.2 are satisfied and conclusions (1)-(4) follow.

In Theorem 10.3, conclusion (3) ensures that the sequence $\{x_m(t)\}$ obtained from (10.9) converges to the solution $x^*(t)$ of (9.1), (2.4). However, in practical evaluation this sequence is approximated by the computed sequence, say, $\{y_m(t)\}$. To find $y_{m+1}(t)$ the function $f$ is approximated by $f_m$. Therefore, the computed sequence $\{y_m(t)\}$ satisfies the recurrence relation

(10.13)     $$y_{m+1}(t) = P_{n-1}(t) + \int_{a_1}^{a_r} g_2(t,s) f_m(s, \underline{y}_m(s)) ds$$

$$y_0(t) = x_0(t) = \bar{x}(t); \; m = 0,1,\ldots \; .$$

With respect to $f_m$, we shall assume the following condition :

Condition $C_1$  For $y_m(t)$, $y_m'(t)$,..., $y_m^{(q)}(t)$ obtained from (10.13), the following inequality is satisfied

$$(10.14) \quad \max_{a_1 \leq t < a_r} \left| f(t,\underline{y}_m(t)) - f_m(t,\underline{y}_m(t)) \right| \leq \Delta \max_{a_1 \leq t \leq a_r} \left| f(t,\underline{y}_m(t)) \right|.$$

Inequality (10.14) corresponds to the relative error in approximating f by $f_m$ for the (m+1)th iteration.

Theorem 10.5 [1]  With respect to (9.1), (2.4) we assume that there exists an approximate solution $\bar{x}(t)$ and the Condition $C_1$ is satisfied. Further, we assume that

(i)    condition (i) of Theorem 10.3

(ii)   $\theta_1 = (1+\Delta)\theta < 1$

(iii)  $N_1 = (1-\theta_1)^{-1}(\varepsilon+\delta+\Delta F)C_{n,0}(a_r-a_1)^n \leq N$

where $F = \max_{a_1 \leq t < a_r} \left| f(t,\bar{x}(t)) \right|$.

Then, the following hold

(1)    all the conclusions (1)-(5) of Theorem 10.3 are valid

(2)    the sequence $\{y_m(t)\}$ obtained from (10.13) remains in $\bar{S}(\bar{x},N_1)$

(3)    the sequence $\{y_m(t)\}$ converges to x*(t), the solution of (9.1), (2.4) if and only if

$$(10.15) \quad \lim_{m \to \infty} \left\| y_{m+1}(t) - P_{n-1}(t) - \int_{a_1}^{a_r} g_2(t,s)f(s,\underline{y}_m(s))ds \right\| = 0$$

(10.16) (4)  $\|x^* - y_{m+1}\| \le (1-\theta)^{-1}[\theta \|y_{m+1} - y_m\|$

$$+ \Delta C_{n,0}(a_r-a_1)^n \max_{a_1 \le t \le a_r} |f(t,\underline{y}_m(t))|].$$

Proof.  Since $\theta_1 < 1$ implies $\theta < 1$ and obviously $N_0 \le N_1$, conditions of Theorem 10.3 are satisfied and conclusion (1) follows.

To prove (2), we note that $\bar{x}(t) \in \bar{S}(\bar{x},N_1)$ and from (10.6) and (10.13), we find

$$y_1(t) - \bar{x}(t) = P_{n-1}(t) - \bar{P}_{n-1}(t) + \int_{a_1}^{a_r} g_2(t,s)[f_0(s,\bar{\underline{x}}(s)) - f(s,\bar{\underline{x}}(s))-\eta(s)]ds.$$

Thus, from Theorem 8.3, we get

$$|y_1^{(j)}(t) - \bar{x}^{(j)}(t)| \le (\varepsilon+\delta)C_{n,j}(a_r-a_1)^{n-j} + C_{n,j}(a_r-a_1)^{n-j}\Delta F, \ 0 \le j \le q$$

and hence

$$\|y_1 - \bar{x}\| \le (\varepsilon+\delta+\Delta F)C_{n,0}(a_r-a_1)^n \le N_1.$$

Next, we assume $y_m(t) \in \bar{S}(\bar{x},N_1)$ and show that $y_{m+1}(t) \in \bar{S}(\bar{x},N_1)$.  From (10.6) and (10.13), we have

$$y_{m+1}(t) - \bar{x}(t) = P_{n-1}(t) - \bar{P}_{n-1}(t) + \int_{a_1}^{a_r} g_2(t,s)[f_m(s,\underline{y}_m(s)) - f(s,\bar{\underline{x}}(s))-\eta(s)]ds$$

and Theorem 8.3 provides

$$|y_{m+1}^{(j)}(t) - \bar{x}^{(j)}(t)| \le (\varepsilon+\delta)C_{n,j}(a_r-a_1)^{n-j} + C_{n,j}(a_r-a_1)^{n-j} \times$$

$$\max_{a_1 \le t \le a_r} [|f_m(t,\underline{y}_m(t)) - f(t,\underline{y}_m(t))|$$

$$+ |f(t,\underline{y}_m(t)) - f(t,\bar{\underline{x}}(t))|]$$

$$\leq C_{n,j}(a_r-a_1)^{n-j}[\epsilon + \delta + \Delta F + (1+\Delta) \max_{a_1 \leq t \leq a_r} |f(t,\underline{y}_m(t)) - f(t,\underline{\bar{x}}(t))|]$$

$$\leq C_{n,j}(a_r-a_1)^{n-j}[\epsilon + \delta + \Delta F + (1+\Delta) \max_{a_1 \leq t \leq a_r} \sum_{i=0}^{q} L_i |y_m^{(i)}(t) - \bar{x}^{(i)}(t)|]$$

$$\leq C_{n,j}(a_r-a_1)^{n-j}[\epsilon + \delta + \Delta F + (1+\Delta) \sum_{i=0}^{q} L_i (C_{n,i}/C_{n,0}(a_r-a_1)^i) \|y_m-\bar{x}\|],$$

$$0 \leq j \leq q.$$

Hence, we get

$$\frac{C_{n,0}(a_r-a_1)^j}{C_{n,j}} |y_{m+1}^{(j)}(t) - \bar{x}^{(j)}(t)| \leq (\epsilon + \delta + \Delta F)C_{n,0}(a_r-a_1)^n$$

$$+ \theta_1 \|y_m - \bar{x}\| , \ 0 \leq j \leq q$$

which gives

$$\|y_{m+1} - \bar{x}\| \leq (1 - \theta_1)N_1 + \theta_1 N_1$$

$$= N_1 .$$

This completes the proof of (2).

From the definitions of $x_{m+1}(t)$ and $y_{m+1}(t)$, we have

$$x_{m+1}(t) - y_{m+1}(t) = P_{n-1}(t) + \int_{a_1}^{a_r} g_2(t,s)f(s,\underline{y}_m(s))ds - y_{m+1}(t)$$

$$+ \int_{a_1}^{a_r} g_2(t,s)[f(s,\underline{x}_m(s)) - f(s,\underline{y}_m(s))]ds$$

and hence, as earlier we find

$$\|x_{m+1} - y_{m+1}\| \leq \|y_{m+1}(t) - P_{n-1}(t) - \int_{a_1}^{a_r} g_2(t,s)f(s,\underline{y}_m(s))ds\|$$

$$+ \theta \|x_m - y_m\| .$$

Since $x_0(t) = y_0(t)$, the above inequality provides

$$(10.17) \qquad \|x_{m+1} - y_{m+1}\| \le \sum_{i=0}^{m} \theta^{m-i} \|y_{i+1}(t) - P_{n-1}(t)$$

$$- \int_{a_1}^{a_r} g_2(t,s)f(s,\underline{y}_i(s))ds\| .$$

Using (10.17) in the triangle inequality, we obtain

$$(10.18) \qquad \|x^* - y_{m+1}\| \le \sum_{i=0}^{m} \theta^{m-i} \|y_{i+1}(t) - P_{n-1}(t)$$

$$- \int_{a_1}^{a_r} g_2(t,s)f(s,\underline{y}_i(s))ds\| + \|x_{m+1} - x^*\| .$$

In (10.18), Theorem 10.3 ensures that $\lim_{m \to \infty} \|x_{m+1} - x^*\| = 0$. Thus, condition (10.15) is necessary and sufficient for the convergence of the sequence $\{y_m(t)\}$ to $x^*(t)$ follows from Toeplitz's lemma "for any $0 \le \alpha < 1$, let $s_m = \sum_{i=0}^{m} \alpha^{m-i} d_i$; $m = 0, 1, \ldots$, then $\lim_{m \to \infty} s_m = 0$ if and only if $\lim_{m \to \infty} d_m = 0$".

Finally, to prove (4) we note that

$$x^*(t) - y_{m+1}(t) = \int_{a_1}^{a_r} g_2(t,s)[f(s,\underline{x}^*(s)) - f(s,\underline{y}_m(s))$$

$$+ f(s,\underline{y}_m(s)) - f_m(s,\underline{y}_m(s))]ds$$

and as earlier, we find

$$(10.19) \qquad \|x^* - y_{m+1}\| \le \theta \|x^* - y_m\| + \Delta C_{n,0}(a_r - a_1)^n \max_{a_1 \le t \le a_r} |f(t,\underline{y}_m(t))| .$$

From (10.19), the inequality (10.16) is obvious.

Remark 10.3   If $\Delta < 1$, then in Theorem 10.5, $N_1$ can be taken as

$$N_1 = (1-\theta_1)^{-1}(\varepsilon + \delta + \Delta(1-\Delta)^{-1} \max_{a_1 \leq t \leq a_r} |f_0(t,\bar{x}(t))|)C_{n,0}(a_r-a_1)^n$$

and the error bound (10.16) can be replaced by

$$\|x^* - y_{m+1}\| \leq (1-\theta)^{-1}[\theta \|y_{m+1} - y_m\|$$

$$+ \Delta(1-\Delta)^{-1} C_{n,0}(a_r-a_1)^n \max_{a_1 \leq t \leq a_r} |f_m(t,\underline{y}_m(t))|].$$

In the next result, we shall assume

Condition $C_2$   For $y_m(t), y_m'(t),\ldots, y_m^{(q)}(t)$ obtained from (10.13), the following inequality is satisfied

$$(10.20) \qquad \max_{a_1 \leq t \leq a_r} |f(t,\underline{y}_m(t)) - f_m(t,\underline{y}_m(t))| \leq \Delta_1 .$$

Inequality (10.20) corresponds to an absolute error in approximating f by $f_m$ for the (m+1)th iteration.

Theorem 10.6 [1]   With respect to (9.1), (2.4) we assume that there exists an approximate solution $\bar{x}(t)$ and the Condition $C_2$ is satisfied. Further, we assume that

(i)   condition (i) of Theorem 10.3

(ii)   condition (ii) of Theorem 10.3

(iii) $N_2 = (1-\theta)^{-1}(\varepsilon + \delta + \Delta_1)C_{n,0}(a_r-a_1)^n \leq N$.

Then, the following hold

(1)   all the conclusions (1)-(5) of Theorem 10.3 are valid

(2)   the sequence $\{y_m(t)\}$ obtained from (10.13) remains in $\bar{S}(\bar{x},N_2)$

(3)   the sequence $\{y_m(t)\}$ converges to $x^*(t)$, the solution of (9.1), (2.4) if and only if (10.15) holds

(4)   $\|x^* - y_{m+1}\| \leq (1-\theta)^{-1} [\theta \|y_{m+1} - y_m\| + \Delta_1 C_{n,0} (a_r - a_1)^n].$

**Proof.** The proof is similar to that of Theorem 10.5.

**Example 10.1 [1]**   For the BVP (1.1), we take $\bar{x}(t) \equiv 0$, then $\varepsilon = 0$, $\delta = |\lambda|$, $D_3 = \{x_0 : |x_0| \leq N\}$, $L_0 = |\alpha\lambda| e^{|\alpha|N}$, $x_1(t) = \frac{\lambda}{2} t(t-1)$ and $\|x_1 - x_0\| = \frac{|\lambda|}{8}$. Thus, the conditions of Theorem 10.3 are satisfied provided

(10.21)        $\theta = \frac{1}{8} |\alpha\lambda| e^{|\alpha|N} < 1$

(10.22)        $N_0 = (1 - \frac{1}{8} |\alpha\lambda| e^{|\alpha|N})^{-1} \frac{|\lambda|}{8} \leq N.$

If $\lambda = 0$, then $N_0 = 0$ and both (10.21), (10.22) are satisfied for all $\alpha$ and $N$. Thus, there exists a solution $x^*(t)$ in $\bar{S}(0,0)$ which must be identically zero, i.e., $x^*(t) \equiv 0$ and it is unique in $\bar{S}(0,\infty)$.

If $\alpha = 0$, then $N_0 = \frac{|\lambda|}{8}$ and the conditions (10.21), (10.22) are satisfied if $\frac{|\lambda|}{8} \leq N$. Thus, there exists a solution $x^*(t)$ in $\bar{S}(0, \frac{|\lambda|}{8})$ which is unique in $\bar{S}(0,N)$ for all $N \geq \frac{|\lambda|}{8}$.

If $\alpha\lambda \neq 0$, then (10.21) and (10.22) both are satisfied if

(10.23)        $|\alpha\lambda| < \dfrac{8p}{1 + pe^p}$

where $p = |\alpha|N$.

In inequality (10.23) the right side attends its maximum at $p = 0.70346\ldots$ which is $2.3240428\ldots$. Thus, we conclude that the BVP (1.1) has a solution in $\bar{S}(0,N_0)$ which is unique in $\bar{S}(0,N)$ provided

(10.24) $\qquad\qquad\qquad |\alpha\lambda| < 2.3240428\ldots$ .

For a given $\alpha$ and $\lambda$ satisfying the inequality (10.24), we find two values of $p(=|\alpha|N)$ so that the inequality (10.23) is satisfied. For example, let $|\alpha| = |\lambda| = 1$, then the two values of $p = N$, are $0.14614\ldots$ and $2.0154\ldots$ . Hence, for this particular case the BVP (1.1) has a solution in $\bar{S}(0, 0.14614\ldots)$ which is unique in $\bar{S}(0, 2.0154\ldots)$. For this solution $x^*(t)$, we find $|x^*(t)| \leq 0.14614\ldots$ .

Example 10.2 [2]  For the BVP

(10.25) $\qquad\qquad\qquad x^{(4)} = x \sin x + \sin t$

(10.26) $\qquad\qquad\qquad x(0) = x'(0) = x(\pi/2) = x'(\pi/2) = 0$

we take $\bar{x}(t) \equiv 0$, so that $\varepsilon = 0$, $\delta = \max\limits_{0 \leq t \leq \pi/2} |\sin t| = 1$, $D_3 = \{x_0 : |x_0| \leq N\}$ and $L_0 = 1 + N$. Thus, the conditions of Theorem 10.3 are satisfied provided

(10.27) $\qquad\qquad\qquad \theta = \dfrac{1}{384} (1+N)(\dfrac{\pi}{2})^4 < 1$

(10.28) $\qquad\qquad\qquad (1-\theta)^{-1} \dfrac{1}{384} (\dfrac{\pi}{2})^4 \leq N.$

Both (10.27) and (10.28) are satisfied if and only if

$$\frac{1}{384} (\frac{\pi}{2})^4 \leq \frac{N}{1 + N(1+N)} \; ,$$

i.e., as long as $0.0161139\ldots = N_1^* \leq N \leq N_2^* = 62.05808\ldots$ .

Since $x_0(t) \equiv 0$, (10.9) provides

$$x_1(t) = \int_0^{\pi/2} g_2(t,s) \sin s \, ds$$

$$= \sin t + (2 - \frac{\pi}{2})(\frac{2}{\pi})^3 t^3 + (\pi-3)(\frac{2}{\pi})^2 t^2 - t$$

and hence $\|x_1 - x_0\| \leq \frac{1}{384}$ .

Thus, if we take $N = N_1^* = 0.0161139$, then $\theta = 0.016109819$ and $N_0 = 0.002646806$. Hence, Theorem 10.3 ensures that (10.25), (10.26) has a unique solution $x^*(t)$ such that $|x^*(t)| \leq 0.0161139$, and the iterative scheme

$$x_{m+1}(t) = \int_0^{\pi/2} g_2(t,s)[x_m(s)\sin x_m(s) + \sin s] ds$$

$$x_0(t) = 0; \quad m = 0,1,\ldots$$

converges to $x^*(t)$ with

$$\|x^* - x_m\| \leq (0.016109819)^m (0.002646806).$$

## COMMENTS AND BIBLIOGRAPHY

Although the important feature of Theorems 10.5 and 10.6 lies in the fact that these results reduce to Theorem 10.3 when $\Delta = 0$ and $\Delta_1 = 0$ respectively, it will be interesting to obtain similar results when the approximating function $f_m(t, \underline{y}_m(t))$ satisfies other error criteria.

1.  Agarwal, R. P. and Loi, S. L. "On approximate Picard's iterates for multipoint boundary value problems", Nonlinear Analysis : Theory, Methods and Appl. 8, 381-391 (1984).
2.  Agarwal, R. P. and Chow, Y. M. "Iterative methods for a fourth order boundary value problem", J. Comp. Appl. Math. 10, 203-217 (1984).
3.  Rall, L. B. Computational Solutions of Nonlinear Operator Equations, John Wiley, New York, 1969.

# 11. QUASILINEARIZATION AND APPROXIMATE QUASILINEARIZATION

It is well recognised that quasilinearization is a fruitful practical method to construct the solution of nonlinear problems in an iterative way. Although the technique as originally developed by Bellman
and Kalaba [5] was motivated by dynamic programming, it is not necessary
to know or to employ dynamic programming to use quasilinear method.

The applications of quasilinear method even for more general problems than (9.1), (2.2) abound in literature [5,7], however, in general
explicit a priori conditions for its convergence to the solution of the
problem are not known. Here, we shall follow the notations and definition of the previous section to provide upper estimates on the length of
the interval $(a_r - a_1)$ so that the sequence $\{x_m(t)\}$ generated by the quasilinear iterative scheme

$$(11.1) \quad x_{m+1}^{(n)}(t) = f(t,\underline{x}_m(t)) + \sum_{i=0}^{q} (x_{m+1}^{(i)}(t) - x_m^{(i)}(t)) \frac{\partial}{\partial x_m^{(i)}(t)} f(t,\underline{x}_m(t))$$

$$(11.2) \quad x_{m+1}^{(j)}(a_i) = A_{j+1,i} ; \quad 1 \le i \le r, \quad 0 \le j \le k_i$$

$$m = 0,1,\ldots$$

with $x_0(t) = \bar{x}(t)$, converges to the unique solution $x^*(t)$ of the BVP
(9.1), (2.4).

Theorem 11.1 [1]  With respect to (9.1), (2.4) we assume that there
exists an approximate solution $\bar{x}(t)$ and

   (i)   the function $f(t,x_0,x_1,\ldots,x_q)$ is continuously differentiable
         with respect to all $x_i$, $0 \le i \le q$ on $[a_1,a_r] \times D_3$

   (ii)  there exist $L_i$, $0 \le i \le q$ nonnegative constants such that
         for all $(t,x_0,x_1,\ldots,x_q) \in [a_1,a_r] \times D_3$

$$\left|\frac{\partial}{\partial x_i} f(t,x_0,x_1,\ldots,x_q)\right| \le L_i$$

(iii)   $3\theta < 1$

(iv)   $N_3 = (1 - 3\theta)^{-1}(\varepsilon + \delta)C_{n,0}(a_r - a_1)^n \le N.$

Then, the following hold

(1)   the sequence $\{x_m(t)\}$ generated by the process (11.1), (11.2) remains in $\bar{S}(\bar{x},N_3)$

(2)   the sequence $\{x_m(t)\}$ converges to the unique solution $x^*(t)$ of the BVP (9.1), (2.4)

(3)   a bound on the error is given by

(11.3)   $\quad \|x_m - x^*\| \le (\frac{2\theta}{1-\theta})^m(1 - \frac{2\theta}{1-\theta})^{-1} \|x_1 - \bar{x}\|$

(11.4)   $\quad \le (\frac{2\theta}{1-\theta})^m(1 - \frac{2\theta}{1-\theta})^{-1}(1-\theta)^{-1}(\varepsilon+\delta)C_{n,0}(a_r-a_1)^n .$

Proof.  First, we shall show that the sequence $\{x_m(t)\}$ remains in $\bar{S}(\bar{x},N_3)$. We define an implicit operator T as follows

(11.5)   $(Tx)(t) = P_{n-1}(t) + \int_{a_1}^{a_r} g_2(t,s)[f(s,\underline{x}(s))$

$$+ \sum_{i=0}^{q} ((Tx)^{(i)}(s) - x^{(i)}(s))\frac{\partial}{\partial x^{(i)}(s)} f(s,\underline{x}(s))]ds$$

whose form is patterned on the integral equation representation of (11.1), (11.2).

Since $\bar{x}(t)$ e $\bar{S}(\bar{x},N_3)$, it is sufficient to show that if $x(t)$ e $\bar{S}(\bar{x},N_3)$, then $(Tx)(t)$ e $\bar{S}(\bar{x},N_3)$. For this, if $x(t)$ e $\bar{S}(\bar{x},N_3)$, then $x(t)$ e $D_3$ and from (10.6) and (11.5), we have

$$(Tx)(t) - \bar{x}(t) = P_{n-1}(t) - \bar{P}_{n-1}(t) + \int_{a_1}^{a_r} g_2(t,s)[f(s,\underline{x}(s))$$

$$+ \sum_{i=0}^{q} ((Tx)^{(i)}(s) - x^{(i)}(s))\frac{\partial}{\partial x^{(i)}(s)} f(s,\underline{x}(s))$$

$$- f(s,\bar{\underline{x}}(s)) - \eta(s)]ds.$$

Thus, an application of Theorem 8.3 provides

$$|(Tx)^j(t) - \bar{x}^{(j)}(t)|$$

$$\leq \varepsilon C_{n,j}(a_r - a_1)^{n-j} + C_{n,j}(a_r - a_1)^{n-j} \max_{a_1 \leq t \leq a_r} [|f(t,\underline{x}(t)) - f(t,\bar{\underline{x}}(t))|$$

$$+ \sum_{i=0}^{q} L_i\{|(Tx)^{(i)}(t) - \bar{x}^{(i)}(t)| + |x^{(i)}(t) - \bar{x}^{(i)}(t)|\} + \delta]$$

and hence, we get

$$\frac{C_{n,0}(a_r - a_1)^j}{C_{n,j}} |(Tx)^{(j)}(t) - \bar{x}^{(j)}(t)| \leq (\varepsilon + \delta)C_{n,0}(a_r - a_1)^n$$

$$+ \sum_{i=0}^{q} C_{n,i}L_i(a_r - a_1)^{n-i} [\|Tx - \bar{x}\| + 2\|x - \bar{x}\|] .$$

From the above inequality, we find

$$\|Tx - \bar{x}\| \leq (\varepsilon + \delta)C_{n,0}(a_r - a_1)^n + \theta\|Tx - \bar{x}\| + 2\theta\|x - \bar{x}\|$$

which gives

$$\|Tx - \bar{x}\| \leq (1-\theta)^{-1}[(\varepsilon + \delta)C_{n,0}(a_r - a_1)^n + 2\theta N_3].$$

Thus, $\|Tx - \bar{x}\| \leq N_3$ follows from the definition of $N_3$.

Next, we shall show the convergence of the sequence $\{x_m(t)\}$. From (11.1), (11.2) we have

(11.6) $\quad x_{m+1}(t) - x_m(t) = \int_{a_1}^{a_r} g_2(t,s) [f(s,\underline{x}_m(s)) - f(s,\underline{x}_{m-1}(s))$

$$+ \sum_{i=0}^{q} \{(x_{m+1}^{(i)}(s) - x_m^{(i)}(s)) \frac{\partial}{\partial x_m^{(i)}(s)} f(s,\underline{x}_m(s))$$

$$- (x_m^{(i)}(s) - x_{m-1}^{(i)}(s)) \frac{\partial}{\partial x_{m-1}^{(i)}(s)} f(s,\underline{x}_{m-1}(s))\}] ds$$

thus, from Theorem 8.3 and the fact that $\{x_m(t)\} \subseteq \bar{S}(\bar{x},N_3)$, we get

$$|x_{m+1}^{(j)}(t) - x_m^{(j)}(t)| \leq C_{n,j}(a_r - a_1)^{n-j} \max_{a_1 \leq t \leq a_r} [2 \sum_{i=0}^{q} L_i |x_m^{(i)}(t) - x_{m-1}^{(i)}(t)|$$

$$+ \sum_{i=0}^{q} L_i |x_{m+1}^{(i)}(t) - x_m^{(i)}(t)|]$$

and hence

$$\frac{C_{n,0}(a_r - a_1)^j}{C_{n,j}} |x_{m+1}^{(j)}(t) - x_m^{(j)}(t)| \leq 2\theta \|x_m - x_{m-1}\| + \theta \|x_{m+1} - x_m\|$$

which provides

$$\|x_{m+1} - x_m\| \leq 2\theta \|x_m - x_{m-1}\| + \theta \|x_{m+1} - x_m\|$$

or

$$\|x_{m+1} - x_m\| \leq \frac{2\theta}{1-\theta} \|x_m - x_{m-1}\|$$

and by an easy induction, we get

$$(11.7) \qquad \|x_{m+1} - x_m\| \leq (\frac{2\theta}{1-\theta})^m \|x_1 - \bar{x}\| .$$

Since $3\theta < 1$, inequality (11.7) implies that $\{x_m(t)\}$ is a Cauchy sequence and hence converges to some $x^*(t) \in \bar{S}(\bar{x}, N_3)$. This $x^*(t)$ is the unique solution of (9.1), (2.4) and can easily be verified.

The error bound (11.3) follows from (11.7) and the triangle inequality

$$\|x_{m+p} - x_m\| \leq \|x_{m+p} - x_{m+p-1}\| + \|x_{m+p-1} - x_{m+p-2}\| + \ldots + \|x_{m+1} - x_m\|$$

$$\leq [(\frac{2\theta}{1-\theta})^{m+p-1} + (\frac{2\theta}{1-\theta})^{m+p-2} + \ldots + (\frac{2\theta}{1-\theta})^m] \|x_1 - \bar{x}\|$$

$$\leq (\frac{2\theta}{1-\theta})^m (1 - \frac{2\theta}{1-\theta})^{-1} \|x_1 - \bar{x}\|$$

and now taking $p \to \infty$.

Next, from (10.6), (11.1), (11.2) we have

$$x_1(t) - x_0(t) = P_{n-1}(t) - \bar{P}_{n-1}(t) + \int_{a_1}^{a_r} g_2(t,s) \times$$

$$[\sum_{i=0}^{q} (x_1^{(i)}(s) - x_0^{(i)}(s))\frac{\partial}{\partial x_0^{(i)}(s)} f(s, \underline{x}_0(s)) - \eta(s)]ds$$

and as earlier, we find

$$(11.8) \qquad \|x_1 - x_0\| \leq (1 - \theta)^{-1}(\epsilon + \delta)C_{n,0}(a_r - a_1)^n.$$

Using (11.8) in (11.3) the inequality (11.4) follows.

Theorem 11.2 [1] Let the conditions of Theorem 11.1 be satisfied. Further, let $f(t, x_0, x_1, \ldots, x_q)$ be continuously twice differentiable with respect to all $x_i$, $0 \leq i \leq q$ on $[a_1, a_r] \times D_3$ and for all

$(t, x_0, x_1, \ldots, x_q) \in [a_1, a_r] \times D_3$

$$\left| \frac{\partial^2}{\partial x_i \partial x_j} f(t, x_0, x_1, \ldots, x_q) \right| \leq L_i L_j K, \quad 0 \leq i, j \leq q.$$

Then,

(11.9)
$$\|x_{m+1} - x_m\| \leq \alpha \|x_m - x_{m-1}\|^2 \leq \frac{1}{\alpha} (\alpha \|x_1 - x_0\|)^{2^m}$$

$$\leq \frac{1}{\alpha} \left[ \frac{1}{2} K (\varepsilon + \delta) \left( \frac{\theta}{1-\theta} \right)^2 \right]^{2^m}$$

where $\alpha = K\theta^2 / 2(1 - \theta) C_{n,0} (a_r - a_1)^n$. Thus, the convergence is quadratic

if $\frac{1}{2} K (\varepsilon + \delta) \left( \frac{\theta}{1-\theta} \right)^2 < 1$.

Proof. From $\{x_m(t)\} \subseteq \bar{S}(\bar{x}, N_3)$ it follows that for all $m$, $x_m(t) \in D_3$.
Further, since $f$ is twice continuously differentiable, we have

(11.10) $f(t, \underline{x}_m(t))$

$$= f(t, \underline{x}_{m-1}(t)) + \sum_{i=0}^{q} (x_m^{(i)}(t) - x_{m-1}^{(i)}(t)) \frac{\partial}{\partial x_{m-1}^{(i)}(t)} f(t, \underline{x}_{m-1}(t))$$

$$+ \frac{1}{2} \left[ \sum_{i=0}^{q} (x_m^{(i)}(t) - x_{m-1}^{(i)}(t)) \frac{\partial}{\partial p_i(t)} \right]^2 f(t, p_0(t), p_1(t), \ldots, p_q(t))$$

where $p_i(t)$ lies between $x_{m-1}^{(i)}(t)$ and $x_m^{(i)}(t)$, $0 \leq i \leq q$.

Using (11.10) in (11.6), we get

$$x_{m+1}(t) - x_m(t)$$

$$= \int_{a_1}^{a_r} g_2(t,s) \left\{ \sum_{i=0}^{q} (x_{m+1}^{(i)}(s) - x_m^{(i)}(s)) \frac{\partial}{\partial x_m^{(i)}(s)} f(s, \underline{x}_m(s)) \right.$$

$$\left. + \frac{1}{2} [\sum_{i=0}^{q} (x_m^{(i)}(s) - x_{m-1}^{(i)}(s)) \frac{\partial}{\partial p_i(s)}]^2 f(s, p_0(s), p_1(s), \ldots, p_q(s)) \right\} ds.$$

Thus, Theorem 8.3 provides

$$|x_{m+1}^{(j)}(t) - x_m^{(j)}(t)|$$

$$\leq C_{n,j}(a_r - a_1)^{n-j} \left\{ \sum_{i=0}^{q} L_i \frac{C_{n,i}}{C_{n,0}} (a_r - a_1)^{-i} \|x_{m+1} - x_m\| \right.$$

$$\left. + \frac{1}{2} [\sum_{i=0}^{q} L_i \frac{C_{n,i}}{C_{n,0}} (a_r - a_1)^{-i}]^2 K \|x_m - x_{m-1}\|^2 \right\}$$

and hence

$$\|x_{m+1} - x_m\| \leq \theta \|x_{m+1} - x_m\| + \frac{K\theta^2}{2C_{n,0}(a_r - a_1)^n} \|x_m - x_{m-1}\|^2$$

which is same as the first part of the inequality (11.9). The second part of (11.9) follows by an easy induction. Finally, the last part is an application of (11.8).

In Theorem 11.1, the conclusion (3) ensures that the sequence $\{x_m(t)\}$ generated from (11.1), (11.2) converges linearly to the unique solution $x^*(t)$ of the BVP (9.1), (2.4). Theorem 11.2 provides sufficient conditions for its quadratic convergence. However, in practical evaluation this sequence is approximated by the computed sequence, say, $\{y_m(t)\}$ which satisfies the recurrence relation

$$(11.11) \qquad y_{m+1}^{(n)}(t) = f_m(t, \underline{y}_m(t)) + \sum_{i=0}^{q} (y_{m+1}^{(i)}(t) - y_m^{(i)}(t)) \frac{\partial}{\partial y_m^{(i)}(t)} f_m(t, \underline{y}_m(t))$$

$$(11.12) \qquad y_{m+1}^{(j)}(a_i) = A_{j+1,i} \; ; \; 1 \le i \le r, \; 0 \le j \le k_i$$

$$m = 0,1,\ldots$$

where $y_0(t) = x_0(t) = \bar{x}(t)$.

With respect to $f_m$, we shall assume the following :

Condition $C_3$ (i) The function $f_m(t,x_0,x_1,\ldots,x_q)$ is continuously differentiable with respect to all $x_i$, $0 \le i \le q$ on $[a_1,a_r] \times D_3$ and for all $(t,x_0,x_1,\ldots,x_q) \in [a_1,a_r] \times D_3$

$$\left| \frac{\partial}{\partial x_i} f_m(t,x_0,x_1,\ldots,x_q) \right| \le L_i$$

(ii) Condition $C_1$ is satisfied.

Theorem 11.3 [1] With respect to (9.1), (2.4) we assume that there exists an approximate solution $\bar{x}(t)$ and the Condition $C_3$ is satisfied. Further, we assume

(i)   conditions (i) and (ii) of Theorem 11.1

(ii)   $\theta_2 = (3 + \Delta)\theta < 1$

(iii) $N_4 = (1 - \theta_2)^{-1}(\epsilon + \delta + \Delta F)C_{n,0}(a_r - a_1)^n \le N$

where $F = \max\limits_{a_1 \le t \le a_r} |f(t,\bar{x}(t))|$.

Then, the following hold

(1)   all the conclusions (1)-(3) of Theorem 11.1 are valid

(2)   the sequence $\{y_m(t)\}$ obtained from (11.11), (11.12) remains in $\bar{S}(\bar{x},N_4)$

(3)   the sequence $\{y_m(t)\}$ converges to $x^*(t)$ the solution of (9.1), (2.4) if and only if $\lim\limits_{m \to \infty} z_m = 0$, where

$$z_m = \|y_{m+1}(t) - P_{n-1}(t) - \int_{a_1}^{a_r} g_2(t,s)f(s,\underline{y}_m(s))ds\|$$

(4)    a bound on the error is given by

(11.13)    $\|x^* - y_{m+1}\| \leq (1 - \theta)^{-1}[2\theta \|y_{m+1} - y_m\|$

$$+ \Delta C_{n,0}(a_r-a_1)^n \max_{a_1 \leq t \leq a_r} |f(t,\underline{y}_m(t))|].$$

Proof. Since $\theta_2 < 1$ implies $3\theta < 1$ and obviously $N_3 \leq N_4$, the conditions of Theorem 11.1 are satisfied and part (1) follows.

To prove (2), we note that $\bar{x}(t) \in \bar{S}(\bar{x},N_4)$ and from (10.6), (11.11), (11.12) we have

$$y_1(t) - \bar{x}(t) = P_{n-1}(t) - \bar{P}_{n-1}(t) + \int_{a_1}^{a_r} g_2(t,s)[f_0(s,\underline{y}_0(s))$$

$$+ \sum_{i=0}^{q} (y_1^{(i)}(s)-y_0^{(i)}(s)) \frac{\partial}{\partial y_0^{(i)}(s)} f_0(s,\underline{y}_0(s))$$

$$- f(s,\underline{y}_0(s)) - \eta(s)]ds$$

and Theorem 8.3 provides

$$\|y_1 - \bar{x}\| \leq (\varepsilon + \delta + \Delta F)C_{n,0}(a_r-a_1)^n + \theta\|y_1 - y_0\|$$

and hence

(11.14)    $\|y_1 - \bar{x}\| \leq (1 - \theta)^{-1}(\varepsilon + \delta + \Delta F)C_{n,0}(a_r-a_1)^n$

$$\leq N_4.$$

Thus, $y_1(t) \in \bar{S}(\bar{x},N_4)$. Next, we assume that $y_m(t) \in \bar{S}(\bar{x},N_4)$ and show that $y_{m+1}(t) \in \bar{S}(\bar{x},N_4)$. From (10.6), (11.11), (11.12) we have

$$y_{m+1}(t) - \bar{x}(t) = P_{n-1}(t) - \bar{P}_{n-1}(t) + \int_{a_1}^{a_r} g_2(t,s)[f_m(s,\underline{y}_m(s))$$

$$+ \sum_{i=0}^{q} (y_{m+1}^{(i)}(s) - y_m^{(i)}(s))\frac{\partial}{\partial y_m^{(i)}(s)} f_m(s,\underline{y}_m(s))$$

$$- f(s,\underline{y}_0(s)) - \eta(s)]ds$$

and from Theorem 8.3, we get

$$|y_{m+1}^{(j)}(t) - \bar{x}^{(j)}(t)| \le (\varepsilon + \delta)C_{n,j}(a_r-a_1)^{n-j}$$

$$+ C_{n,j}(a_r-a_1)^{n-j} \max_{a_1 \le t \le a_r} [\sum_{i=0}^{q} L_i |y_{m+1}^{(i)}(t)-y_m^{(i)}(t)|$$

$$+ (1+\Delta)|f(t,\underline{y}_m(t))-f(t,\underline{y}_0(t))| + \Delta|f(t,\underline{y}_0(t))|]$$

and hence, we find

$$\|y_{m+1} - \bar{x}\| \le (\varepsilon + \delta + \Delta F)C_{n,0}(a_r-a_1)^n$$

$$+ \theta\|y_{m+1} - y_m\| + (1+\Delta)\theta\|y_m - y_0\|$$

$$\le (\varepsilon + \delta + \Delta F)C_{n,0}(a_r-a_1)^n$$

$$+ (2+\Delta)\theta\|y_m - y_0\| + \theta\|y_{m+1} - y_0\| .$$

From the last inequality, we obtain

$$\|y_{m+1} - \bar{x}\| \le (1-\theta)^{-1}[(\varepsilon+\delta+\Delta F)C_{n,0}(a_r-a_1)^n + (2+\Delta)\theta N_4]$$

$$= N_4 .$$

This completes the proof of part (2).

Next, from the definitions of $x_{m+1}(t)$ and $y_{m+1}(t)$, we have

$$x_{m+1}(t) - y_{m+1}(t) = P_{n-1}(t) + \int_{a_1}^{a_r} g_2(t,s) f(s,\underline{y}_m(s)) ds - y_{m+1}(t)$$

$$+ \int_{a_1}^{a_r} g_2(t,s) [f(s,\underline{x}_m(s)) - f(s,\underline{y}_m(s))$$

$$+ \sum_{i=0}^{q} (x_{m+1}^{(i)}(s) - x_m^{(i)}(s)) \frac{\partial}{\partial x_m^{(i)}(s)} f(s,\underline{x}_m(s))] ds$$

and hence as earlier, we find

(11.15) $\qquad \|x_{m+1} - y_{m+1}\| \le z_m + \theta \|x_m - y_m\| + \theta \|x_{m+1} - x_m\|$ .

Using (11.7) in (11.15), we get

$$\|x_{m+1} - y_{m+1}\| \le z_m + \theta \|x_m - y_m\| + \theta (\frac{2\theta}{1-\theta})^m \|x_1 - \bar{x}\| .$$

Since $x_0(t) = y_0(t) = \bar{x}(t)$, the above inequality provides

(11.16) $\qquad \|x_{m+1} - y_{m+1}\| \le \sum_{i=0}^{m} \theta^{m-i} [z_i + \theta (\frac{2\theta}{1-\theta})^i \|x_1 - \bar{x}\|]$ .

Using (11.16) in the triangle inequality, we obtain

(11.17) $\qquad \|y_{m+1} - x^*\| \le \|x_{m+1} - x^*\| + \sum_{i=0}^{m} \theta^{m-i} [z_i + \theta (\frac{2\theta}{1-\theta})^i \|x_1 - \bar{x}\|]$.

In (11.17), Theorem 11.1 ensures that $\lim_{m \to \infty} \|x_{m+1} - x^*\| = 0$. Thus, from the Toeplitz lemma $\lim_{m \to \infty} \|y_{m+1} - x^*\| = 0$ if and only if $\lim_{m \to \infty} [z_m + \theta (\frac{2\theta}{1-\theta})^m \|x_1 - \bar{x}\|] = 0$, however, $\lim_{m \to \infty} (\frac{2\theta}{1-\theta})^m = 0$, and hence if and only if $\lim_{m \to \infty} z_m = 0$.

To prove (4), we note that

$$x^*(t) - y_{m+1}(t) = \int_{a_1}^{a_r} g_2(t,s)[f(s,x^*(s)) - f(s,\underline{y}_m(s)) + f(s,\underline{y}_m(s))$$

$$-f_m(s,\underline{y}_m(s)) - \sum_{i=0}^{q} (y_{m+1}^{(i)}(s) - y_m^{(i)}(s))\frac{\partial}{\partial y_m^{(i)}(s)}f_m(s,\underline{y}_m(s))]ds$$

and hence

$$\|x^* - y_{m+1}\| \le \theta\|x^* - y_m\| + \theta\|y_{m+1} - y_m\|$$

$$+ \Delta C_{n,0}(a_r - a_1)^n \max_{a_1 \le t < a_r} |f(t,\underline{y}_m(t))|$$

$$\le 2\theta\|y_{m+1} - y_m\| + \Delta C_{n,0}(a_r - a_1)^n \max_{a_1 \le t < a_r} |f(t,\underline{y}_m(t))|$$

$$+ \theta\|x^* - y_{m+1}\|$$

which is same as (11.13).

Theorem 11.4 [1]  Let the conditions of Theorem 11.3 be satisfied. Further, let $f_m = f_0$ for all $m = 1,2,\ldots$ and $f_0(t,x_0,x_1,\ldots,x_q)$ be continuously twice differentiable with respect to all $x_i$, $0 \le i \le q$ on $[a_1,a_r] \times D_3$ and for all $(t,x_0,x_1,\ldots,x_q) \in [a_1,a_r] \times D_3$

$$\left|\frac{\partial^2}{\partial x_i \partial x_j} f_0(t,x_0,x_1,\ldots,x_q)\right| \le L_i L_j K, \quad 0 \le i, j \le q.$$

Then,

(11.18)  $\|y_{m+1} - y_m\| \le \alpha\|y_m - y_{m-1}\|^2 \le \frac{1}{\alpha}(\alpha\|y_1 - y_0\|)^{2^m} \le \frac{1}{\alpha}[\frac{1}{2}K(\varepsilon + \delta + \Delta F)(\frac{\theta}{1-\theta})^2]^{2^m}$

where $\alpha$ is same as in Theorem 11.2.

Proof. As in the proof of Theorem 11.2, we have

$$y_{m+1}(t) - y_m(t)$$

$$= \int_{a_1}^{a_r} g_2(t,s) \left\{ \sum_{i=0}^{q} (y_{m+1}^{(i)}(s) - y_m^{(i)}(s)) \frac{\partial}{\partial y_m^{(i)}(s)} f_0(s, \underline{y}_m(s)) \right.$$

$$\left. + \frac{1}{2} [\sum_{i=0}^{q} (y_m^{(i)}(s) - y_{m-1}^{(i)}(s)) \frac{\partial}{\partial p_i(s)}]^2 f_0 (s, p_0(s), p_1(s), \ldots, p_q(s)) \right\} ds$$

where $p_i(t)$ lies between $y_{m-1}^{(i)}(t)$ and $y_m^{(i)}(t)$, $0 \le i \le q$.

Thus, as earlier we get

$$\|y_{m+1} - y_m\| \le \theta \|y_{m+1} - y_m\| + (1-\theta) \alpha \|y_m - y_{m-1}\|^2$$

which is same as the first part of (11.18). The last part of (11.18) follows from (11.14).

Remark 11.1 Results similar to Theorems 11.1-11.4 for the other BVPs can easily be stated.

Example 11.1 [1] Once again, we consider the BVP (1.1). As in Example 10.1 the conditions of Theorem 11.1 in $D_3 = \{x_0 : |x_0| \le N\}$ are satisfied provided

(11.19) $$3\theta = \frac{3}{8} |\alpha\lambda| e^{|\alpha|N} < 1$$

(11.20) $$N_3 = (1 - \frac{3}{8} |\alpha\lambda| e^{|\alpha|N})^{-1} \frac{|\lambda|}{8} \le N.$$

We note that both the conditions (11.19), (11.20) are satisfied if

(11.21) $$|\alpha\lambda| < \frac{8p}{1 + 3p\, e^p}$$

where $p = |\alpha|N$.

In inequality (11.21) the right side attends its maximum at $p = 0.4589622...$ which is $1.550476...$ . Thus, we conclude that the quasilinear scheme (11.1), (11.2) for (1.1) converges to the solution provided

$$|\alpha\lambda| < 1.550476... .$$

From this value of $p$, we find

$$\theta = 0.19780386 \; |\alpha\lambda|$$

and hence, for $|\alpha\lambda| = 1$, inequality (11.3) provides

$$\|x_m - x^*\| \leq (0.49315586)^m (1.972993118)\frac{|\lambda|}{8} .$$

Next, we shall apply Theorem 11.2, for this we note that $K = \frac{1}{|\lambda|} e^{-|\alpha|N}$. Hence, from the above obtained values

$$\frac{1}{2} K(\varepsilon + \delta) (\frac{\theta}{1-\theta})^2 = \frac{1}{2} e^{-p} (\frac{\theta}{1-\theta})^2 < 1$$

is obvious. Thus, the convergence is indeed quadratic.

Example 11.2 [2]  As in Example 10.2 the conditions of Theorem 11.1 in $D_3 = \{x_0 : |x_0| \leq N\}$ for the BVP (10.25), (10.26) are satisfied provided

(11.22)
$$3\theta = \frac{3}{384} (1 + N) (\frac{\pi}{2})^4 < 1$$

(11.23)
$$N_3 = (1 - 3\theta)^{-1} \frac{1}{384} (\frac{\pi}{2})^4 \leq N.$$

The inequalities (11.22), (11.23) are satisfied if and only if

$$\frac{1}{384} (\frac{\pi}{2})^4 \leq \frac{N}{1 + 3N(1 + N)} ,$$

i.e., as long as $0.016659943... \le N \le 20.00807171...$ . If we take $N = 0.016659943$, then $3\theta = 0.048355429$ and $N_3 = 0.016659942$.

Thus, Theorem 11.1 guarantees the convergence of the quasilinear scheme

$$(11.24) \quad x_{m+1}(t) = \int_0^{\pi/2} g_2(t,s)\{x_m(s)\sin x_m(s) + \sin s + (x_{m+1}(s) - x_m(s)) \times$$

$$[x_m(s)\cos x_m(s) + \sin x_m(s)]\} ds$$

$$x_0(t) = 0; \quad m = 0,1,...$$

to the unique solution $x^*(t)$ of the BVP (10.25), (10.26). Also, (11.3) provides

$$\|x^* - x_m\| \le (0.032765075)^m (0.016659942).$$

Finally, we note that in Theorem 11.2, $K = \dfrac{(2+N)}{(1+N)^2}$ . Thus,

$\alpha = \dfrac{1}{2} \dfrac{(2+N)}{(1+N)^2} (\dfrac{\theta}{1-\theta})^2$ which is in fact less than $2.61826... \times 10^{-4}$ and

hence (11.24) indeed converges quadratically.

Example 11.3 [3]  Following the Remark 11.1 it is easy to verify that the iterative scheme

$$(11.25) \quad x_{m+1}^{(4)}(t) = t \cos x_m(t) + \cos t + (x_{m+1}(t) - x_m(t))(-t \sin x_m(t))$$

$$(11.26) \quad x_{m+1}(-\tfrac{1}{2}) = x_{m+1}'(-\tfrac{1}{2}) = x_{m+1}''(\tfrac{1}{2}) = x_{m+1}'''(\tfrac{1}{2}) = 0$$

$$m = 0,1,...$$

with $x_0(t) \equiv 0$ converges quadratically to the unique solution $x^*(t)$ of the BVP

$$x^{(4)} = t \cos x + \cos t$$

$$x(-\tfrac{1}{2}) = x'(-\tfrac{1}{2}) = x''(\tfrac{1}{2}) = x'''(\tfrac{1}{2}) = 0.$$

The numerical results for (11.25), (11.26) by using forward-forward and backward-backward methods are presented in Tables 4 and 5 respectively.

Example 11.4 [3]   The iterative scheme

$$(11.27) \qquad x_{m+1}^{(4)}(t) = \tfrac{8}{81} x_m^5(t) + (x_{m+1}(t) - x_m(t))(\tfrac{40}{81} x_m^4(t))$$

$$(11.28) \qquad x_{m+1}(1) = \tfrac{1}{3}, \ x'_{m+1}(1) = -\tfrac{1}{3}, \ x''_{m+1}(2) = \tfrac{1}{12}, \ x'''_{m+1}(2) = -\tfrac{1}{8}$$

$$m = 0,1,\ldots$$

with $x_0(t) = \tfrac{1}{48}(38 - 29t + 8t^2 - t^3)$ converges quadratically to the unique solution $x^*(t)$ of the BVP

$$x^{(4)} = \tfrac{8}{81} x^5$$

$$x(1) = \tfrac{1}{3}, \ x'(1) = -\tfrac{1}{3}, \ x''(2) = \tfrac{1}{12}, \ x'''(2) = -\tfrac{1}{8}.$$

The numerical results for (11.27), (11.28) by using forward-forward and backward-backward methods are presented in Tables 6 and 7 respectively. We also take the initial approximation $x_0(t) = \tfrac{1}{3t}$ and present the numerical results in Tables 8 and 9.

In all numerical integration of differential equations we use fourth order Runge-Kutta method with $h = \tfrac{1}{256}$ and the necessary intermediate values are interpolated using nearest four points.

Table 4.

| t | $x_1(t)$ | $x_2(t)$ | $x_3(t)$ |
|---|---|---|---|
| -0.500 | 0.00000000 D 00 | 0.00000000 D 00 | 0.00000000 D 00 |
| -0.375 | 0.40808726 D-02 | 0.40753516 D-02 | 0.40753517 D-02 |
| -0.250 | 0.15134278 D-01 | 0.15113285 D-01 | 0.15113285 D-01 |
| -0.125 | 0.31483371 D-01 | 0.31438591 D-01 | 0.31438592 D-01 |
| 0.000 | 0.51625410 D-01 | 0.51550175 D-01 | 0.51550177 D-01 |
| 0.125 | 0.74268027 D-01 | 0.74157326 D-01 | 0.74157327 D-01 |
| 0.250 | 0.98361724 D-01 | 0.98212189 D-01 | 0.98212191 D-01 |
| 0.375 | 0.12312857 D 00 | 0.12293837 D 00 | 0.12293838 D 00 |
| 0.500 | 0.14808711 D 00 | 0.14785566 D 00 | 0.14785566 D 00 |

Table 5.

| t | $x_1(t)$ | $x_2(t)$ | $x_3(t)$ |
|---|---|---|---|
| -0.500 | 0.22865781 D-15 | -0.19997418 D-06 | -0.19964519 D-06 |
| -0.375 | 0.40962931 D-02 | 0.40905433 D-02 | 0.40905433 D-02 |
| -0.250 | 0.15190086 D-01 | 0.15168632 D-01 | 0.15168631 D-01 |
| -0.125 | 0.31597001 D-01 | 0.31551361 D-01 | 0.31551360 D-01 |
| 0.000 | 0.51808350 D-01 | 0.51731727 D-01 | 0.51731725 D-01 |
| 0.125 | 0.74527251 D-01 | 0.74414536 D-01 | 0.74414534 D-01 |
| 0.250 | 0.98700959 D-01 | 0.98548725 D-01 | 0.98548721 D-01 |
| 0.375 | 0.12354941 D 00 | 0.12335580 D 00 | 0.12335579 D 00 |
| 0.500 | 0.14858998 D 00 | 0.14835438 D 00 | 0.14835438 D 00 |

Table 6.

| t | $x_1(t)$ | $x_2(t)$ |
|---|---|---|
| 1.000 | 0.33333333 D 00 | 0.33333333 D 00 |
| 1.125 | 0.29325366 D 00 | 0.29325366 D 00 |
| 1.250 | 0.25618515 D 00 | 0.25618515 D 00 |
| 1.375 | 0.22188361 D 00 | 0.22188361 D 00 |
| 1.500 | 0.19010487 D 00 | 0.19010487 D 00 |
| 1.625 | 0.16060479 D 00 | 0.16060479 D 00 |
| 1.750 | 0.13313922 D 00 | 0.13313922 D 00 |
| 1.875 | 0.10746401 D 00 | 0.10746401 D 00 |
| 2.000 | 0.83335022 D-01 | 0.83335022 D-01 |

Table 7.

| t | $x_1(t)$ | $x_2(t)$ |
|---|---|---|
| 1.000 | 0.33333333 D 00 | 0.33333333 D 00 |
| 1.125 | 0.29325366 D 00 | 0.29325366 D 00 |
| 1.250 | 0.25618515 D 00 | 0.25618515 D 00 |
| 1.375 | 0.22188362 D 00 | 0.22188362 D 00 |
| 1.500 | 0.19010489 D 00 | 0.19010489 D 00 |
| 1.625 | 0.16060481 D 00 | 0.16060481 D 00 |
| 1.750 | 0.13313924 D 00 | 0.13313924 D 00 |
| 1.875 | 0.10746403 D 00 | 0.10746403 D 00 |
| 2.000 | 0.83335049 D-01 | 0.83335049 D-01 |

Table 8.

| t | $x_1(t)$ | $x_2(t)$ | $x_3(t)$ |
|---|---|---|---|
| 1.000 | 0.33333333 D 00 | 0.33333333 D 00 | 0.33333333 D 00 |
| 1.125 | 0.29325361 D 00 | 0.29325366 D 00 | 0.29325366 D 00 |
| 1.250 | 0.25618496 D 00 | 0.25618515 D 00 | 0.25618515 D 00 |
| 1.375 | 0.22188321 D 00 | 0.22188361 D 00 | 0.22188361 D 00 |
| 1.500 | 0.19010421 D 00 | 0.19010487 D 00 | 0.19010487 D 00 |
| 1.625 | 0.16060383 D 00 | 0.16060479 D 00 | 0.16060479 D 00 |
| 1.750 | 0.13313794 D 00 | 0.13313922 D 00 | 0.13313922 D 00 |
| 1.875 | 0.10746240 D 00 | 0.10746401 D 00 | 0.10746401 D 00 |
| 2.000 | 0.83333088 D-01 | 0.83335022 D-01 | 0.83335022 D-01 |

Table 9.

| t | $x_1(t)$ | $x_2(t)$ | $x_3(t)$ |
|---|---|---|---|
| 1.000 | 0.33333333 D 00 | 0.33333333 D 00 | 0.33333333 D 00 |
| 1.125 | 0.29325361 D 00 | 0.29325366 D 00 | 0.29325366 D 00 |
| 1.250 | 0.25618497 D 00 | 0.25618515 D 00 | 0.25618515 D 00 |
| 1.375 | 0.22188323 D 00 | 0.22188362 D 00 | 0.22188362 D 00 |
| 1.500 | 0.19010425 D 00 | 0.19010489 D 00 | 0.19010489 D 00 |
| 1.625 | 0.16060388 D 00 | 0.16060481 D 00 | 0.16060481 D 00 |
| 1.750 | 0.13313801 D 00 | 0.13313924 D 00 | 0.13313924 D 00 |
| 1.875 | 0.10746249 D 00 | 0.10746403 D 00 | 0.10746403 D 00 |
| 2.000 | 0.83333187 D-01 | 0.83335049 D-01 | 0.83335049 D-01 |

## COMMENTS AND BIBLIOGRAPHY

It is known [8] that Chaplygin's method [6] is a variant of
Newton's method, which itself is labelled as Quasilinearization when
applied to nonlinear BVPs [4,5,7]. While in the present section we
have succeeded in establishing the convergence of the quasilinear
methods for the higher order equations, the monotonic convergence property shared by second order problems [5,7] needs further investigations.

1.  Agarwal, R. P. "Quasilinearization and approximate quasilineariza-
    tion for multipoint boundary value problems", J. Math. Anal. Appl.
    107, 317-330 (1985).
2.  Agarwal, R. P. and Chow, Y. M. "Iterative methods for a fourth
    order boundary value problem", J. Comp. Appl. Math. 10, 203-217
    (1984).
3.  Agarwal, R. P. and Usmani, R. A. "Iterative methods for solving
    right focal point boundary value problems", J. Comp. Appl. Math.
    to appear.
4.  Antosiewicz, H. A. "Newton's method and boundary value problems",
    J. Comput. System Sci. 2, 177-202 (1968).
5.  Bellman, R. E. and Kalaba, R. E. Quasilinearization and Nonlinear
    Boundary Value Problems, Elsevier, New York, 1965.
6.  Chaplygin, S. A. Collected Papers on Mechanics and Mathematics,
    Moscow, 1954.
7.  Lee, E. S. Quasilinearization and Invariant Imbedding, Academic
    Press, New York, 1968.
8.  Vidossich, G. "Chaplygin's method is Newton's method", J. Math.
    Anal. Appl. 66, 188-206 (1978).

## 12. BEST POSSIBLE RESULTS : WEIGHT FUNCTION TECHNIQUE

Here, we shall employ weight function technique of Weissinger [15] and Collatz [10] to show that the region of existence and uniqueness of solutions of BVPs can be enlarged as compared to what can be deduced from the results of sections 9 and 10. More so, in some special cases where the explicit form of the Green's function and some of its properties are known, it is possible to find best existence and uniqueness intervals. For this, we need

**Definition 12.1** A system of nonnegative continuous functions (weight functions) $w_j(t)$, $0 \leq j \leq q$ in $[a_1, a_r]$ is called admissible with respect to the Green's function $g_2(t,s)$ if

(i) there exist smallest positive constants $k_j$, $0 \leq j \leq q$ such that

(12.1) $$\phi_j(t) = \int_{a_1}^{a_r} |g_2^{(j)}(t,s)| ds \leq k_j w_j(t); \quad a_1 \leq t \leq a_r, \quad 0 \leq j \leq q$$

(ii) there exist finite smallest positive constants $\bar{k}_{ij}$, $0 \leq i$, $j \leq q$ such that

(12.2) $$\sup_{a_1 \leq t \leq a_r} \frac{\int_{a_1}^{a_r} |g_2^{(j)}(t,s)| w_i(s) ds}{w_j(t)} \leq \bar{k}_{ij}, \quad 0 \leq i, j \leq q.$$

**Definition 12.2** If we denote $K_i = \max\{\bar{k}_{ij}, 0 \leq j \leq q\}$, $0 \leq i \leq q$ then the inequalities (12.2) can be written as

(12.3) $$\sup_{a_1 \leq t \leq a_r} \frac{\int_{a_1}^{a_r} |g_2^{(j)}(t,s)| w_i(s) ds}{w_j(t)} \leq K_i, \quad 0 \leq i, j \leq q.$$

The system of these constants $k_j$, $K_j$, $0 \leq j \leq q$ is called the associated system of constants for the admissible system $w_j(t)$, $0 \le j \le q$.

The class of admissible functions with respect to the Green's function $g_2(t,s)$ is nonempty. For example, an obvious choice is $w_j(t) \equiv 1$, $0 \le j \le q$. Another possibility is

$$(12.4) \qquad w_j(t) = \phi_j(t) = \int_{a_1}^{a_r} |g_2^{(j)}(t,s)| ds, \ 0 \le j \le q.$$

For (12.4) inequalities (12.1) are obvious with $k_j = 1$, $0 \le j \le q$. Further, since

$$\sup_{a_1 \le t \le a_r} \frac{\int_{a_1}^{a_r} |g_2^{(j)}(t,s)| |w_i(s) ds}{w_j(t)} \le \sup_{a_1 \le t \le a_r} \int_{a_1}^{a_r} |g_2^{(i)}(t,s)| ds$$

$$= \max_{a_1 \le t \le a_r} \phi_i(t), \ 0 \le i, j \le q$$

there exist finite smallest positive constants $\bar{k}_{ij}$, $0 \le i, j \le q$ such that inequalities (12.2) are satisfied.

For a given admissible system $w_j(t)$, $0 \le j \le q$ we shall consider the Banach space of all functions $x(t)$ e $B = C^{(q)}[a_1, a_r]$ with the finite norm

$$(12.5) \qquad \|x\| = \max_{0 \le j \le q} \left\{ \sup_{a_1 \le t \le a_r} \frac{|x^{(j)}(t)|}{w_j(t)} \right\}.$$

The space B for the admissible system $w_j(t) \equiv 1$, $0 \le j \le q$ is obviously nonempty. Further, for the system defined in (12.4), we consider the set M of n times continuously differentiable functions satisfying (9.10). This set $M \subseteq C^{(q)}[a_1, a_r]$ and if $x(t)$ e M, then it can be written as

$$x(t) = \int_{a_1}^{a_r} g_2(t,s) x^{(n)}(s) ds$$

and hence

$$\frac{|x^{(j)}(t)|}{w_j(t)} \leq \frac{1}{w_j(t)} \int_{a_1}^{a_r} |g_2^{(j)}(t,s)| |x^{(n)}(s)| ds$$

$$\leq \max_{a_1 \leq t < a_r} |x^{(n)}(t)|, \quad 0 \leq j \leq q$$

which implies that $\|x\|$ is finite.

Theorem 12.1 [2]  Let $w_j(t)$, $0 \leq j \leq q$ be an admissible system with respect to the Green's function $g_2(t,s)$ and $k_j$, $K_j$, $0 \leq j \leq q$ be the associated system of constants.  Further, we assume that the function f satisfies the Lipschitz condition (9.12) on $[a_1, a_r] \times D_4$, where

$$D_4 = \{(x_0, x_1, \ldots, x_q) : |x_j - P_{n-1}^{(j)}(t)| \leq k^*(1-\nu)^{-1}(L+c)w_j(t), \quad 0 \leq j \leq q\}$$

and

$$k^* = \max_{0 \leq j \leq q} k_j$$

$$L = \max_{a_1 \leq t < a_r} |f(t,0,0,\ldots,0)|$$

$$c = \max_{a_1 \leq t < a_r} \sum_{i=0}^{q} L_i |P_{n-1}^{(i)}(t)|$$

(12.6) $$\nu = \sum_{i=0}^{q} L_i K_i < 1.$$

Then, the following hold

(1)    there exists a unique solution $x^*(t)$ of (9.1), (2.4) in
$\bar{S}(P_{n-1}, k^*(1-\nu)^{-1}(L+c)) \subseteq B$

(2)    the sequence $\{x_m(t)\}$ defined by (10.9) with $x_0(t) = P_{n-1}(t)$
converges to $x^*(t)$ and

(12.7)    $$\|x^* - x_m\| \leq k^*\nu^m(1-\nu)^{-1}(L+c).$$

**Proof.**    We shall show that the operator $T : \bar{S}(P_{n-1}, k^*(1-\nu)^{-1}(L+c))$
$\to C^{(n)}[a_1, a_r]$ defined in (9.2) satisfies the conditions of Theorem 10.2.
Let $x(t) \in \bar{S}(P_{n-1}, k^*(1-\nu)^{-1}(L+c))$, then from the definition of norm
(12.5), we have

$$\|x - P_{n-1}\| = \max_{0 \leq j \leq q} \left\{ \sup_{a_1 \leq t \leq a_r} \frac{|x^{(j)}(t) - P_{n-1}^{(j)}(t)|}{w_j(t)} \right\} \leq k^*(1-\nu)^{-1}(L+c)$$

which implies that

$$|x^{(j)}(t) - P_{n-1}^{(j)}(t)| \leq k^*(1-\nu)^{-1}(L+c)w_j(t), \quad 0 \leq j \leq q$$

and hence $x(t) \in D_4$.    Further, if $x(t), y(t) \in \bar{S}(P_{n-1}, k^*(1-\nu)^{-1}(L+c))$,
then

$$|(Tx)^{(j)}(t) - (Ty)^{(j)}(t)| \leq \int_{a_1}^{a_r} |g^{(j)}(t,s)| \sum_{i=0}^{q} L_i |x^{(i)}(s) - y^{(i)}(s)| ds$$

$$= \int_{a_1}^{a_r} |g^{(j)}(t,s)| \sum_{i=0}^{q} L_i \frac{|x^{(i)}(s) - y^{(i)}(s)|}{w_i(s)} w_i(s) ds$$

$$\leq \int_{a_1}^{a_r} |g^{(j)}(t,s)| \sum_{i=0}^{q} L_i w_i(s) ds \|x - y\|$$

and hence

$$\frac{|(Tx)^{(j)}(t) - (Ty)^{(j)}(t)|}{w_j(t)} \leq \sum_{i=0}^{q} L_i \frac{\int_{a_1}^{a_r} |g^{(j)}(t,s)| w_i(s) ds}{w_j(t)} \|x - y\|$$

$$\leq \sum_{i=0}^{q} L_i \, \bar{k}_{ij} \, \|x - y\|$$

$$\leq \sum_{i=0}^{q} L_i \, K_i \, \|x - y\|$$

from which it follows that

$$\|Tx - Ty\| \leq \nu \|x - y\| \, .$$

Next, let $x(t) \, \epsilon \, \bar{S}(P_{n-1}, k*(1-\nu)^{-1}(L+c))$, then

$$|(Tx)^{(j)}(t) - P_{n-1}^{(j)}(t)| \leq \int_{a_1}^{a_r} |g^{(j)}(t,s)| |f(s,\underline{x}(s))| ds$$

$$\leq \int_{a_1}^{a_r} |g^{(j)}(t,s)| [L + \sum_{i=0}^{q} L_i \frac{|x^{(i)}(s) - P_{n-1}^{(i)}(s)|}{w^{(i)}(s)} w^{(i)}(s)$$

$$+ \sum_{i=0}^{q} L_i |P_{n-1}^{(i)}(s)|] ds$$

$$\leq \int_{a_1}^{a_r} |g^{(j)}(t,s)| [(L+c) + \sum_{i=0}^{q} L_i w^{(i)}(s) \|x - P_{n-1}\|] ds$$

and hence

$$\frac{|(Tx)^{(j)}(t) - P_{n-1}^{(j)}(t)|}{w_j(t)} \leq k*(L+c) + \nu \|x - P_{n-1}\|$$

or

$$\|Tx - P_{n-1}\| \leq k*(L+c) + k*\nu(1-\nu)^{-1}(L+c)$$

$$= k*(1-\nu)^{-1}(L+c).$$

This completes the proof of our theorem.

Remark 12.1  Results similar to Theorem 12.1 for the other BVPs can easily be stated.

For the admissible system defined in (12.4) the constant k* is immediately available and it is 1, however, the computation of the constants $K_j$, $0 \leq j \leq q$ is not an easy task except for some simple BVPs. For instance :

Example 12.1 [8]  For the BVP (7.1), (7.2)

$$w_0(t) = \frac{1}{2} (t-a_1)(a_2-t)$$

$$w_1(t) = \frac{(t-a_1)^2 + (a_2-t)^2}{2(a_2-a_1)}$$

$$\bar{k}_{00} = \frac{5}{48} (a_2-a_1)^2, \quad \bar{k}_{01} = \frac{\sqrt{3}-1}{4\sqrt{3}} (a_2-a_1)^2$$

$$\bar{k}_{01} = \frac{8}{25} (a_2-a_1), \quad \bar{k}_{11} = \frac{1}{3} (a_2-a_1)$$

and hence

$$K_0 = \frac{\sqrt{3}-1}{4\sqrt{3}} (a_2-a_1)^2, \quad K_1 = \frac{1}{3} (a_2-a_1)$$

Thus, from the definition of $\nu$, we have

(12.8)        $$\nu = \frac{\sqrt{3}-1}{4\sqrt{3}} L_0(a_2-a_1)^2 + \frac{1}{3} L_1(a_2-a_1) < 1$$

which is sharper than (7.4).

Example 12.2 [5]  For the differential equation

(12.9)                    $$x''' = f(t,x)$$

together with

(12.10)          $x(a_1) = A, \quad x'(a_1) = B, \quad x(a_2) = C$

or

(12.11)          $x(a_1) = A, \quad x(a_2) = B, \quad x'(a_2) = C$

we have

$$w_0(t) = \begin{cases} \dfrac{1}{6}(t-a_1)^2(a_2-t) & \text{for (12.10)} \\[2ex] \dfrac{1}{6}(t-a_1)(a_2-t)^2 & \text{for (12.11)} \end{cases}$$

$$K_0 = \bar{k}_{00} = \frac{1}{60}(a_2-a_1)^3$$

and hence

(12.12)          $\nu = \dfrac{1}{60} L_0 (a_2-a_1)^3 < 1$

which is sharper than $\theta = \dfrac{2}{81} L_0 (a_2-a_1)^3 < 1.$

**Example 12.3** [5]  For the differential equation

(12.13)          $x''' = f(t, x, x', x'')$

together with (12.10) or (12.11) an elementary though tedious calculation provides

(12.14)          $\nu = \dfrac{3}{160} L_0 (a_2-a_1)^3 + \dfrac{33}{320} L_1 (a_2-a_1)^2 + \dfrac{3}{8} L_2 (a_2-a_1) < 1$

which is sharper than

(12.15)          $\theta = \dfrac{2}{81} L_0 (a_2-a_1)^3 + \dfrac{1}{6} L_1 (a_2-a_1)^2 + \dfrac{2}{3} L_2 (a_2-a_1) < 1.$

Example 12.4 [3]  For the differential equation (12.9) together with

(12.16) $\qquad x(a_1) = A, \quad x'(a_1) = B, \quad x'(a_2) = C$

or

(12.17) $\qquad x'(a_1) = A, \quad x(a_2) = B, \quad x'(a_2) = C$

we have

$$w_0(t) = \begin{cases} \dfrac{1}{4}\,(a_2-a_1)(t-a_1)^2 - \dfrac{1}{6}\,(t-a_1)^3 & \text{for (12.16)} \\[3mm] \dfrac{1}{4}\,(a_2-a_1)(a_2-t)^2 - \dfrac{1}{6}\,(a_2-t)^3 & \text{for (12.17)} \end{cases}$$

$$K_0 = \bar{k}_{00} = \frac{1}{24}\,(a_2-a_1)^3$$

and hence

(12.18) $\qquad\qquad \nu = \dfrac{1}{24}\,L_0(a_2-a_1)^3 < 1$

which is sharper than $\theta = \dfrac{1}{12}\,L_0(a_2-a_1)^3 < 1$.

Example 12.5 [3]  For each of the BVPs (12.13), (12.16); (12.13), (12.17) an easy calculation provides

(12.19) $\qquad \nu = \dfrac{7}{120}\,L_0(a_2-a_1)^3 + \dfrac{\sqrt{3}-1}{4\sqrt{3}}\,L_1(a_2-a_1)^2 + \dfrac{1}{3}\,L_3(a_2-a_1) < 1$

which is sharper than

(12.20) $\qquad \theta = \dfrac{1}{12}\,L_0(a_2-a_1)^3 + \dfrac{1}{8}\,L_1(a_2-a_1)^2 + \dfrac{1}{2}\,L_2(a_2-a_1) < 1.$

In some particular BVPs it is possible to choose an admissible system $w_j(t)$, $0 \le j \le q$ so that the existence and uniqueness interval is best possible. For instance :

Example 12.6 [6]  We shall consider the BVP (9.1), (2.7) :

Let $z(t)$ be the solution of the initial value problem

(12.21) $$D_n z(t) = z^{(n)} + \sum_{i=0}^{q} c_i z^{(i)} = 0$$

(12.22)
$$z^{(i)}(0) = 0, \quad 0 \le i \le n-2$$
$$z^{(n-1)}(0) = 1$$

and $t_i(\bar{c})$ be the first positive constant such that $z^{(i)}(t_i(\bar{c})) = 0$, i.e., $z^{(i)}(t) > 0$ for all $t \in (0, t_i(\bar{c}))$, where $\bar{c}$ represents the vector $(c_0, c_1, \ldots, c_q)$ appearing in (12.21).

Lemma 12.2  Let $x(t), y(t) \in C^{(n)}[a_1, a_2]$ and

(12.23) $$D_n x(t) \ge 0$$

(12.24) $$D_n y(t) \le 0, \quad t \in (a_1, a_2]$$

$$x^{(i)}(a_1) = y^{(i)}(a_1), \quad 0 \le i \le n-1.$$

Then,

(12.25) $$x^{(i)}(t) \ge y^{(i)}(t), \quad t \in [a_1, a_1 + t_i(\bar{c})].$$

Proof.  Inequalities (12.23) and (12.24) with some $\phi_1(t), \phi_2(t) \ge 0$, $t \in [a_1, a_2]$ can be written as

$$D_n x(t) - \phi_1(t) = 0$$

$$D_n y(t) + \phi_2(t) = 0.$$

Hence, if $x_0(t)$ is such that $D_n x_0(t) = 0$, $x_0^{(i)}(a_1) = x^{(i)}(a_1) = y^{(i)}(a_1)$, $0 \le i \le n-1$ then

$$x(t) = x_0(t) + \int_{a_1}^{t} z(t-s)\phi_1(s)ds$$

$$y(t) = x_0(t) - \int_{a_1}^{t} z(t-s)\phi_2(s)ds.$$

Thus, we have

$$x^{(i)}(t) - y^{(i)}(t) = \int_{a_1}^{t} \frac{\partial^i z(t-s)}{\partial t^i} [\phi_1(s) + \phi_2(s)]ds.$$

But, since $\dfrac{\partial^i z(t-s)}{\partial t^i} \geq 0$ as long as $a \leq s \leq t \leq a + t_i(\bar{c})$ the inequalities (12.25) follow.

Lemma 12.3   Let the vector $\bar{c} \geq 0$ and $q \leq p$. Then, for any $\lambda > 1$, $t_p(\bar{c}) > t_p(\lambda\bar{c})$.

Proof.   Let $y(t)$ be such that

(12.26) $$y^{(n)}(t) + \lambda \sum_{i=0}^{q} c_i y^{(i)}(t) = 0$$

satisfying (12.22).  Then, since $q \leq p$ it is necessary that for all $t \in [a_1, a_1 + t_p(\lambda\bar{c})]$, $y^{(i)}(t) \geq 0$, $0 \leq i \leq q$ and hence

$$y^{(n)}(t) + \sum_{i=0}^{q} c_i y^{(i)}(t) \leq 0.$$

From Lemma 12.2, we find that $z^{(p)}(t) \geq y^{(p)}(t)$, $t \in [a_1, a_1 + t_p(\bar{c})]$. Thus, $y^{(p)}(t)$ must vanish before $a_1 + t_p(\bar{c})$.

Remark 12.2   As a consequence of Lemma 12.3, if $(a_2 - a_1) < t_p(\bar{c})$, then there exists a constant $\lambda > 1$ such that $(a_2 - a_1) = t_p(\lambda\bar{c})$. Further, for

this choice of $\lambda$ the BVP (12.26), (3.20) has a nontrivial solution
$y(t)$ such that $y^{(p)}(t) > 0$, $t \in (a_1, a_2)$. To fix the choice of $y(t)$
we require that $y^{(n-1)}(a_1) = 1$.

Lemma 12.4  Let the vector $\bar{c} \geq 0$, $q \leq p$, $(a_2 - a_1) < t_p(\bar{c})$ and $\lambda$, $y(t)$
be as in Remark 12.2.  Then, for the system $w_j(t) = y^{(j)}(t)$, $0 \leq j \leq q$
the following hold

(i)
$$\frac{\int_{a_1}^{a_2} |g_5^{(j)}(t,s)| \sum_{i=0}^{q} c_i w_i(s) ds}{w_j(t)} = \frac{1}{\lambda} \, , \, 0 \leq j \leq q$$

(ii)   there exists a smallest possible constant $k^*$ such that

$$\int_{a_1}^{a_2} |g_5^{(j)}(t,s)| ds \leq \frac{k^*}{\lambda} w_j(t), \, 0 \leq j \leq q.$$

Proof.   Since $y(t)$ is a solution of (12.26), (3.20) it can be represented as

$$y(t) = \lambda \int_{a_1}^{a_2} [-g_5(t,s)] \sum_{i=0}^{q} c_i \, y^{(i)}(s) ds$$

and hence from Lemma 3.6, we find

$$w_j(t) = \lambda \int_{a_1}^{a_2} |g_5^{(j)}(t,s)| \sum_{i=0}^{q} c_i \, w_i(s) ds \, , \, 0 \leq j \leq q.$$

Remark 12.3   As a consequence of Lemma 12.4, this system $w_j(t) = y^{(j)}(t)$,
$0 \leq j \leq q$ is an admissible system with respect to the Green's function
$g_5(t,s)$.

Theorem 12.5  Let $q \leq p$, $(a_2 - a_1) < t_p(\bar{L})$ and $w_j(t)$, $0 \leq j \leq q$ defined
in Lemma 12.4 be taken as the admissible system with respect to the

Green's function $g_5(t,s)$. Further, we assume that the function f satisfies the Lipschitz condition (9.12) on $[a_1,a_2] \times D_5$, where

$$D_5 = \{(x_0,x_1,\ldots,x_q) : |x_j - P_{n-1}^{(j)}(t)| \le \frac{k^*}{\lambda} (1 - \frac{1}{\lambda})^{-1}(L+c)w_j(t), \ 0 \le j \le q\}$$

and $\lambda$ is as in Remark 12.2, $k^*$ is as in Lemma 12.4, L and c are as in Theorem 12.1. Then, the following hold

(1)　there exists a unique solution $x^*(t)$ of (9.1), (2.7) in

$$\bar{S}(P_{n-1}, \frac{k^*}{\lambda} (1 - \frac{1}{\lambda})^{-1}(L + c)) \subseteq B$$

(2)　the sequence $\{x_m(t)\}$ defined by (10.9) with $x_0(t) = P_{n-1}(t)$ converges to $x^*(t)$ and

$$\|x^* - x_m\| \le \frac{k^*}{\lambda} (\frac{1}{\lambda})^m(1 - \frac{1}{\lambda})^{-1}(L + c) .$$

Proof.　The proof is similar to that of Theorem 12.1.

Remark 12.4　The inequality $(a_2-a_1) < t_p(\bar{L})$ is best possible, i.e., it cannot be replaced by the equality. Indeed in case of equality the BVP (12.26), (2.7) with $A_i = 0$, $0 \le i \le n-2$ and $B \ne 0$ has no solution, and the BVP (12.26), (3.20) has trivial as well as nontrivial solutions.

Corollary 12.6　Let f satisfy the Lipschitz condition (9.12) on $[a_1,a_2] \times R^{q+1}$ and $(a_2-a_1) < t_p(\bar{L})$. Then, the BVP (9.1), (2.7) has a unique solution.

In [4], $t_p(\bar{L})$ has been computed explicitly for several particular cases. For example, when n = 3 and q = p = 0 then $t_0(L_0)$ is the first positive root of the equation

$$2 \sin(\frac{\sqrt{3}}{2} L_0^{1/3} b - \frac{\pi}{6}) + \exp(- \frac{3}{2} L_0^{1/3} b ) = 0$$

which justifies some of the statements made in Example 1.4.

Example 12.7 [7]   Consider the BVP

(12.27) $$x^{(4)} = f(t,x)$$

(12.28) $$x(-1) = x'(-1) = x(1) = x'(1) = 0 :$$

Lemma 12.7   For the differential equation

$$x^{(4)} = \mu x$$

together with (12.28), the following hold

(1)  $\mu \leq 0$ is not an eigenvalue

(2)  $\mu = \mu_n > 0$ is an eigenvalue provided $\cosh 2\mu_n^{1/4} \cos 2\mu_n^{1/4} = 1$

(3)  the eigenfunction $x_1(t)$ corresponding to the first eigenvalue $\mu_1$ is of fixed sign in $(-1,1)$ and can be expressed as

$$x_1(t) = \cosh \mu_1^{1/4}(t+1)\cos \mu_1^{1/4} + \sinh \mu_1^{1/4}(t+1)\sin \mu_1^{1/4} - \cos \mu_1^{1/4} t$$

(4)
$$\frac{\displaystyle\int_{-1}^{1} |g_2(t,s)||x_1(s)|ds}{|x_1(t)|} = \frac{1}{\mu_1} \ , \ \mu_1 = 31.28524\ldots$$

(5)  there exists a smallest possible constant $k^*$ such that

$$\int_{-1}^{1} |g_2(t,s)|ds \leq k^*|x_1(t)|.$$

Proof.  Let $\mu^* < 0$ be an eigenvalue and $x_*(t)$ be the corresponding eigenfunction, then $x_*(t)x_*^{(4)}(t) = \mu^*[x_*(t)]^2$ and integration by parts provides

$$\int_{-1}^{1} |x_*^{(2)}(t)|^2 \, dt = \mu^* \int_{-1}^{1} |x_*(t)|^2 \, dt$$

which implies $x_*^{(2)}(t) = 0$ almost everywhere.  But, then $x_*(t) = 0$

follows from the boundary conditions (12.28).  The proof of other parts is by direct computation.

**Theorem 12.8**    Let $L_0 < \mu_1$ and $w_0(t) = |x_1(t)|$ defined in Lemma 12.7 be taken as the admissible function with respect to the Green's function $g_2(t,s)$.  Further, we assume that the function $f$ satisfies the Lipschitz condition (9.12) on $[-1,1] \times D_6$, where

$$D_6 = \{x_0 : |x_0| \leq k^*(1 - \frac{L_0}{\mu_1})^{-1} L w_0(t)\}$$

and $k^*$ is as in Lemma 12.7 and $L$ as in Theorem 12.1.  Then, the following hold

(1)   there exists a unique solution $x^*(t)$ of (12.27), (12.28) in

$$\bar{S}(0, k^*(1 - \frac{L_0}{\mu_1})^{-1} L)) \subseteq B$$

(2)   the sequence $\{x_m(t)\}$ defined by (10.9) with $x_0(t) \equiv 0$ converges to $x^*(t)$ and

$$\|x^* - x_m\| \leq k^*(\frac{L_0}{\mu_1})^m (1 - \frac{L_0}{\mu_1})^{-1} L .$$

**Proof.**  The proof is similar to that of Theorem 12.1.

**Example 12.8** [8]    Consider the BVP (9.4), (9.5) :

If $u$, $v \in R^m$, we say $u \leq v$ if and only if $u_i \leq v_i$, $1 \leq i \leq m$.

**Definition 12.3**    Let $E$ be a real vector space.  A generalized norm for $E$ is a mapping $\|\cdot\|_G : E \to R_+^m$ denoted by $\|u\|_G = (\alpha_1(u),\ldots,\alpha_m(u))$ such that

(i)    $\|u\|_G \geq 0$, that is, $\alpha_i(u) \geq 0$ for all $i$

(ii)   $\|u\|_G = 0$ if and only if $u = 0$, that is, $\alpha_i(u) = 0$ for all $i$ if and only if $u = 0$

(iii)  $\|\lambda u\|_G = |\lambda| \|u\|_G$, that is, $\alpha_i(\lambda u) = |\lambda|\alpha_i(u)$

(iv) $\quad \|u + v\|_G \leq \|u\|_G + \|v\|_G$, that is, $\alpha_i(u+v) \leq \alpha_i(u) + \alpha_i(v)$.

The space $(E, \|\cdot\|_G)$ is called a generalized normed space. The topology in this space is given in the following way : For each $u \in E$, and $\varepsilon > 0$, let $B_\varepsilon(u) = \{v \in E : \|v - u\|_G < \varepsilon w\}$, where $w = (1,\ldots,1) \in R^m$. Then, $\{B_\varepsilon(u) : u \in E, \varepsilon > 0\}$ forms a basis for a topology on E. The same topology can be induced by the usual norm $\|\cdot\|$ which is defined as follows : If $\|u\|_G = (\alpha_1(u),\ldots,\alpha_m(u))$, then $\|u\| = \max\{\alpha_1(u),\ldots,\alpha_m(u)\}$. Since the topology of the normed space $(E, \|\cdot\|)$ is given by the basis of neighbourhoods $V_\varepsilon(u) = \{v \in E : \|v - u\| < \varepsilon\}$, $u \in E$, $\varepsilon > 0$ and $V_\varepsilon(u) = B_\varepsilon(u)$, both the above definitions of norm define the same topology on E and are equivalent. Thus, from the topological point of view there is no need for introducing the generalized norm. However, we have more flexibility when working with generalized spaces. This is clear in the case of Contraction mapping theorem which can be stated as follows :

**Theorem 12.9**   Let $(E, \|\cdot\|_G)$ be a complete generalized normed space, and let for $r \in R_+^m$, $r > 0$, $\bar{S}(u^0,r) = \{u \in E : \|u - u^0\|_G \leq r\}$. Let T map $\bar{S}(u^0,r)$ into itself and for all $u,v \in \bar{S}(u^0,r)$

$$\|Tu - Tv\|_G \leq \Delta\|u - v\|_G$$

where $\Delta \geq 0$ in an $m \times m$ matrix with the spectral radius $\rho(\Delta) < 1$. Then, the following hold

(1)   T has a unique fixed point $u^*$ in $\bar{S}(u^0,r)$

(2)   the sequence $\{u^k\}$ defined by

$$u^{k+1} = Tu^k ; \quad k = 0,1,\ldots$$

converges to $u^*$ with

$$\|u^* - u^k\|_G \leq \Delta^k r.$$

**Remark 12.5**   The system of functions

$$w_0(t) = \sin\frac{\pi(t-a_1)}{(a_2-a_1)}$$

and

$$w_1(t) = \frac{2}{(a_2-a_1)} \sin \frac{\pi(t-a_1)}{(a_2-a_1)} + \frac{\pi(a_2-2t+a_1)}{(a_2-a_1)^2} \cos \frac{\pi(t-a_1)}{(a_2-a_1)}$$

is an admissible system with respect to the Green's function $g_0(t,s)$. In fact, an easy computation provides

$$k^* = \frac{1}{2\pi}(a_2-a_1)^2, \quad K_0 = \frac{1}{\pi^2}(a_2-a_1)^2, \quad K_1 = \frac{4}{\pi^2}(a_2-a_1).$$

In the following result we shall consider the complete generalized normed space of all functions $u(t) \in B = (C^{(1)}[a_1,a_2],R^m)$ with the finite generalized norm $\|u\|_G = (\alpha_1(u),\ldots,\alpha_m(u))$, where

$$\alpha_i(u) = \max \left\{ \sup_{a_1 \leq t \leq a_2} \frac{|u_i(t)|}{w_0(t)}, \quad \sup_{a_1 \leq t \leq a_2} \frac{|u_i'(t)|}{w_1(t)} \right\}.$$

**Theorem 12.10** Let $w_0(t)$ and $w_1(t)$ defined in Remark 12.5 be taken as the admissible system with respect to the Green's function $g_0(t,s)$. Further, we assume that for all $(t,u^0,v^0)$, $(t,u^1,v^1) \in [a_1,a_2] \times D_7$

$$|F(t,u^0,v^0) - F(t,u^1,v^1)| \leq \bar{L}_0|u^0 - u^1| + \bar{L}_1|v^0 - v^1|$$

where $\bar{L}_0$ and $\bar{L}_1$ are $m \times m$ nonnegative matrices, and

$$D_7 = \{(u,v) : |u - \ell^1 - \frac{\ell^2 - \ell^1}{a_2-a_1}(t-a_1)| \leq \frac{1}{2\pi}(a_2-a_1)^2(I-\Delta)^{-1}\bar{L}w_0(t),$$

$$|v - \frac{\ell^2 - \ell^1}{a_2-a_1}| \leq \frac{1}{2\pi}(a_2-a_1)^2(I-\Delta)^{-1}\bar{L}w_1(t)\}$$

$$\Delta = \frac{1}{\pi^2}\bar{L}_0(a_2-a_1)^2 + \frac{4}{\pi^2}\bar{L}_1(a_2-a_1)$$

with $\rho(\Delta) < 1$, and $\bar{L} = \max_{a_1 \leq t \leq a_2} |F(t,0,0)|$.

Then, the following hold

    (1) there exists a unique solution $u^*(t)$ of (9.4), (9.5) in

$$\bar{S}(\ell^1 + \frac{\ell^2 - \ell^1}{a_2 - a_1}(t-a_1), \frac{1}{2\pi}(a_2-a_1)^2(I-\Delta)^{-1}\bar{L}) \subseteq B$$

    (2) the sequence $\{u^k(t)\}$ defined by

$$u^{k+1}(t) = \ell^1 + \frac{\ell^2 - \ell^1}{a_2 - a_1}(t-a_1) + \int_{a_1}^{a_2} g_0(t,s)F(s,u^k(s),u'^k(s))ds$$

    with $u^0(t) = \ell^1 + \frac{\ell^2 - \ell^1}{a_2 - a_1}(t-a_1)$ converges to $u^*(t)$ and

$$\|u^* - u^k\|_G \leq \frac{1}{2\pi}(a_2-a_1)^2 \Delta^k(I-\Delta)^{-1}\bar{L} .$$

Proof. The proof is similar to that of Theorem 12.1.

Remark 12.6    The BVP (9.1), (2.9) with $q = 0$, where f satisfies the Lipschitz condition (9.12) on $[a_1, a_2] \times R$ has a unique solution provided

(12.29)          $\dfrac{L_0}{\pi^{2m}}(a_2-a_1)^{2m} < 1.$

In fact this BVP is equivalent to the system

$$u_i'' = u_{i+1}, \quad 1 \leq i \leq m-1$$

$$u_m'' = f(t,u_1)$$

$$u(a_1) = \ell^1, \; u(a_2) = \ell^2$$

and the condition $\rho(\Delta) < 1$, where

$$\Delta = \frac{(a_2-a_1)^2}{\pi^2} \begin{vmatrix} 0 & 1 & 0 & \cdots & 0 \\ 0 & 0 & 1 & \cdots & 0 \\ \cdot & \cdot & \cdot & \cdots & \cdot \\ 0 & 0 & 0 & \cdots & 1 \\ L_0 & 0 & 0 & \cdots & 0 \end{vmatrix}$$

is same as (12.29). This result covers a particular case given in [14].

Remark 12.7    If the function F is independent of u', then the condition $\rho(\Delta) < 1$ is best possible. For this, we note that the uncoupled system

$$u_i'' + L_0 u_i = 0, \quad 1 \le i \le m$$

$$u(a_1) = u(a_2) = 0$$

where $\dfrac{1}{\pi^2} L_0 (a_2 - a_1)^2 = 1$ has infinite number of solutions

$$u_i(t) = c_1 \sin \sqrt{L_0}(t - a_1), \quad 1 \le i \le m$$

where $c_1$ is an arbitrary constant.  Further, the system

$$u_i'' + u_{i+1} = 0, \quad 1 \le i \le m-1$$

$$u_m'' + L_0 u_1 = 0$$

$$u(a_1) = u(a_2) = 0$$

where $\dfrac{1}{\pi^{2m}} L_0 (a_2 - a_1)^{2m} = 1$ has infinite number of solutions

$$u_i(t) = c_2 L_0^{(i-1)/m} \sin L_0^{1/2m}(t - a_1), \quad 1 \le i \le m$$

where $c_2$ is an arbitrary constant.

## COMMENTS AND BIBLIOGRAPHY

It would be desirable to prove the following conjecture : For the admissible system $w_j(t)$, $0 \le j \le q$ defined in (12.4) with respect to the Green's function $g_2(t,s)$ the constants $K_i$ are $\bar{k}_{iq}$, $0 \le i \le q$.  Inequality (12.19) has been improved recently in [1].  It is well recognized

that working with generalized normed spaces for the systems one achieves
much more information about the solutions than what can be inferred by
considering the usual norms. In particular the component-wise study
enlarges the domain of existence and uniqueness of solutions, weakens
the convergence conditions and provides sharper error estimates,e.g.,
see [9,11-13].

1.  Aftabizadeh, A. R. and Wiener, J. "On the solutions of third order
    non-linear boundary value problems", Trends in the Theory and Pra-
    ctice of Non-linear Analysis, Ed. V. Lakshmikantham, Elsevier
    Science Publishers B. V. (North Holland) 1985, 1-6.
2.  Agarwal, R. P. "Boundary value problems for higher order integro-
    differential equations", Nonlinear Analysis : Theory, Methods and
    Appl. 7, 259-270 (1983).
3.  Agarwal, R. P. "Nonlinear two-point boundary value problems",
    Indian J. Pure and Appl. Math. 4, 757-769 (1973).
4.  Agarwal, R. P. "Two-point problems for nonlinear third order
    differential equations", J. Math. Phyl. Sci. 8, 571-576 (1974).
5.  Agarwal, R. P. and Krishnamoorthy, P. R. "On the uniqueness of
    solutions of nonlinear boundary value problems", J. Math. Phyl.
    Sci. 10, 17-31 (1976).
6.  Agarwal, R. P. and Krishnamoorthy, P. R. "Boundary value problems
    for nth order ordinary differential equations", Bull. Inst. Math.
    Acad. Sinica  7, 211-230 (1979).
7.  Agarwal, R. P. and Wilson, S. J. "On a fourth order boundary value
    problem", Utilitas Mathematica  26, 297-310 (1984).
8.  Agarwal, R. P. and Vosmanský , J. "Two-point boundary value pro-
    blems for second order systems", Arch. Math. (Brnö)  19, 1-8
    (1983).
9.  Agarwal, R. P. "Contraction and approximate contraction with an
    application to multi-point boundary value problems", J. Comp.
    Appl. Math. 9, 315-325 (1983).
10. Collatz, L. "Einige Anwendungen functionalanalytischer Methoden
    in der praktischen Analysis", Z. Angew. Math. Phys. 4, 327-357
    (1953).
11. Ortega, J. M. and Rheinboldt, W. C. Iterative Solutions of Non-
    linear Equations in Several Variables, Academic Press, New York,
    1970.
12. Schröder, J. Operator Inequalities, Academic Press, New York,
    1980.
13. Šeda, V. "On a vector multipoint boundary value problem", Arch.
    Math. (Brnö) to appear.
14. Usmani, R. A. "A uniqueness theorem for a boundary value problem",
    Proc. Amer. Math. Soc. 77, 329-335 (1979).
15. Weissinger, J. "Zur Theorie und Anwendung des Iterationsverfahrens",
    Math. Nachr. 8, 193-212 (1952).

## 13. BEST POSSIBLE RESULTS : SHOOTING METHODS

In example 12.6, we employed weight function technique for the BVP (9.1), (2.7) when $q \leq p$ and obtained best possible existence and uniqueness interval in terms of the Lipschitz constants $L_0, \ldots, L_q$. Here, we shall again assume that $q \leq p$, but the function f satisfies the condition

$$(13.1) \quad M_j(x_j - y_j) \leq f(t, x_0, \ldots, x_{j-1}, y_j, x_{j+1}, \ldots, x_q) - f(t, x_0, \ldots, x_j, \ldots, x_q)$$
$$\leq K_j(x_j - y_j)$$

whenever $x_j \geq y_j$, $0 \leq j \leq q$ to obtain best possible existence and uniqueness interval for each of the BVPs (9.1), (2.7); (9.1), (2.8) in terms of $M_j$, $K_j$, $0 \leq j \leq q$. Obviously, the condition (13.1) is equivalent to the Lipschitz condition (9.12), but more informative particularly since there are no sign restrictions on the constants.

**Lemma 13.1** Let $x(t)$, $y(t) \in C^{(n)}[a_1, a_2]$, satisfy (12.23) and (12.24) respectively on $[a_1, a_2]$ and $x^{(i)}(a_2) = y^{(i)}(a_2)$, $0 \leq i \leq n-1$. Then,

$$(-1)^{n+i} x^{(i)}(t) \geq (-1)^{n+i} y^{(i)}(t), \quad t \in [a_2 - z_i(\bar{c}), a_2]$$

where $z_i(\bar{c})$ is the first positive constant such that $z^{(i)}(-z_i(\bar{c})) = 0$, i.e., $(-1)^{n+i+1} z^{(i)}(t) > 0$, $t \in (-z_i(\bar{c}), 0)$ and $z(t)$ is the solution of (12.21), (12.22).

**Proof.** The proof is similar to that of Lemma 12.2.

**Theorem 13.2** Suppose that the condition (13.1) is satisfied. Then,

(1) the BVP (9.1), (2.7) has at most one solution provided $(a_2 - a_1) < t_p(\bar{K})$

(2) for n odd the BVP (9.1), (2.8) has at most one solution provided $(a_2 - a_1) < z_p^*$, where $z_p^* = z_p(M_0, K_1, M_2, K_3, \ldots)$

(3)   for n even the BVP (9.1), (2.8) has at most one solution
      provided $(a_2-a_1) < z_p^{**}$, where $z_p^{**} = z_p(K_0,M_1,K_2,M_3,\ldots)$.

Proof.  We shall prove only (2) and the proof of (1) and (3) follows
similarly.  Let $x_1(t)$ and $x_2(t)$ be two solutions of (9.1), (2.8) then
the function $h(t) = x_1(t) - x_2(t)$ is such that $h^{(i)}(a_2) = 0$, $0 \leq i \leq n-2$
and from the uniqueness of solutions of initial value problems
$h^{(n-1)}(a_2) \neq 0$.  Let $h^{(n-1)}(a_2) > 0$, then there exists a $t^* \in [a_1,a_2)$
such that $(-1)^{n+p+1}h^{(p)}(t) > 0$, $t \in (t^*,a_2)$ and $h^{(p)}(t^*) = 0$.  Thus,
$(-1)^{n+k+1}h^{(k)}(t) \geq 0$; $t \in [t^*,a_2]$, $0 \leq k \leq q$ and hence condition (13.1)
on using the fact that n is odd provides

$$h^{(n)}(t) + \sum_{\substack{j \leq q \\ \text{even}}} M_j h^{(j)}(t) + \sum_{\substack{j \leq q \\ \text{odd}}} K_j h^{(j)}(t) \leq 0.$$

Let $w(t)$ be the solution of the initial value problem

$$w^{(n)}(t) + \sum_{\substack{j \leq q \\ \text{even}}} M_j w^{(j)}(t) + \sum_{\substack{j \leq q \\ \text{odd}}} K_j w^{(j)}(t) = 0$$

$$w^{(i)}(a_2) = 0, \quad 0 \leq i \leq n-2$$

$$w^{(n-1)}(a_2) = h^{(n-1)}(a_2) \neq 0.$$

Then, from Lemma 13.1 we have $(-1)^{n+p} h^{(p)}(t) \leq (-1)^{n+p} w^{(p)}(t) \leq 0$,
$t \in [t^*,a_2]$.  This implies that $w^{(p)}(t)$ vanishes in $(t^*,a_2)$.  However,
since $a_2 - t^* \leq a_2 - a_1 < z_p^*$ and $(-1)^{n+p+1} w^{(p)}(t) > 0$, $t \in (a_2 - z_p^*,a_2)$
we get a contradiction to our assumption.

Lemma 13.3   If $(a_2-a_1) < t_p(\bar{K})$, then the BVP

$$(13.2) \qquad y^{(n)}(t) + \sum_{j=0}^{q} K_j y^{(j)}(t) = f(t,0,\ldots,0)$$

$$y^{(i)}(a_1) = 0, \quad 0 \le i \le n-2$$

(13.3)

$$y^{(p)}(a_2) = m \in R$$

has a unique solution.

**Proof.** It can easily be verified that the only solution of (13.2), (13.3) is

$$y(t) = [y_0^{(p)}(a_2)]^{-1}[m - \int_{a_1}^{a_2} y_0^{(p)}(a_2-s)f(s,0,\ldots,0)ds]y_0(t)$$

$$+ \int_{a_1}^{t} y_0(t-s)f(s,0,\ldots,0)ds$$

where $y_0(t)$ is the solution of the initial value problem

$$y_0^{(n)}(t) + \sum_{j=0}^{q} K_j y_0^{(j)}(t) = 0$$

$$(13.4) \qquad y_0^{(i)}(a_1) = 0, \quad 0 \le i \le n-2$$

$$y_0^{(n-1)}(a_1) = 1.$$

Using this lemma we can take $m_2 > m$ sufficiently large and positive so that $y^{(i)}(t) > 0$ on $(a_1, a_2]$, $0 \le i \le p$ and $y^{(n-1)}(a_1) > 0$. Similarly, we can take $m_1 < m$ sufficiently large but negative so that $y^{(i)}(t) < 0$ on $(a_1, a_2]$, $0 \le i \le p$ and $y^{(n-1)}(a_1) < 0$.

**Lemma 13.4** Suppose that $a_1 < T < a_1 + t_p(\bar{K})$ and on $a_1 < t \le T$ the function $y(t)$ and its derivatives upto order $p$ are positive and satisfy

(13.2); $y^{(i)}(a_1) = 0$, $0 \le i \le n-2$, $y^{(n-1)}(a_1) > 0$. Further, we assume that the condition (13.1) is satisfied. Then, if $x(t)$ is the solution of (9.1) satisfying $x^{(i)}(a_1) = y^{(i)}(a_1)$, $0 \le i \le n-1$

$$(13.5) \qquad y^{(i)}(t) \le x^{(i)}(t); \ t \ \epsilon \ [a_1,T], \ 0 \le i \le p.$$

Proof. Since $x(t)$ and $y(t)$ satisfy the same initial conditions, we have

$$(13.6) \qquad x(t) - y(t) = \int_{a_1}^{t} y_0(t-s)[f(s,x(s),x'(s),\ldots,x^{(q)}(s))$$

$$+ \sum_{j=0}^{q} K_j \, x^{(j)}(s) - f(s,0,\ldots,0)]ds$$

where $y_0(t)$ is the solution of (13.4). However, as long as $x^{(i)}(t) \ge 0$, $0 \le i \le q$, the condition (13.1) provides

$$(13.7) \quad f(s,x(s),x'(s),\ldots,x^{(q)}(s)) - f(s,0,\ldots,0) + \sum_{j=0}^{q} K_j \, x^{(j)}(s) \ge 0.$$

Thus, (13.5) follows from (13.6) as long as $x^{(i)}(t) \ge 0$, $0 \le i \le q$. Since $x^{(n-1)}(a_1) > 0$, $x^{(p)}(t) > 0$ on $(a_1,T_1)$, if $T_1 \ge T$ then obviously $x^{(i)}(t) \ge 0$, $0 \le i \le q$ on $[a_1,T]$ and the proof is complete. If $T_1 < T$, then $x^{(p)}(T_1) = 0$ and from the above considerations, we get $y^{(p)}(T_1) \le x^{(p)}(T_1) = 0$ which is a contradiction to our assumption.

Remark 13.1  The inequality (13.7) is reversed as long as $x^{(i)}(t) \le 0$, $0 \le i \le q$. Thus, if in Lemma 13.4 "positive" is replaced by "negative" and "$y^{(n-1)}(a_1) > 0$" by "$y^{(n-1)}(a_1) < 0$", then the conclusion (13.5) is replaced by

$$(13.8) \qquad y^{(i)}(t) \ge x^{(i)}(t); \ t \ \epsilon \ [a_1,T], \ 0 \le i \le p.$$

**Theorem 13.5**  In the conclusion (1) of Theorem 13.2 "at most one" can be replaced by "has a unique".

**Proof.**  Since the BVP (9.1), (2.7) for $\bar{x}(t) = x(t) - \sum\limits_{i=0}^{n-2} \dfrac{(t-a_1)^i}{i!} A_i$ is exactly of the same type as that for $x(t)$ with $\bar{x}^{(i)}(a_1) = 0$, $0 \le i \le n-2$ we assume for simplicity that $A_i = 0$, $0 \le i \le n-2$.

Let $m_2 > m > B$ be sufficiently large and positive in the sense of Lemma 13.3 so that $y(t,m_2)$ is the unique solution of (13.2), (13.3). As a result of Lemma 13.4 and standard theorem on the continuation of solutions of differential equations the solution $x_2(t)$ of (9.1) satisfying the initial conditions $x_2^{(i)}(a_1) = 0$, $0 \le i \le n-2$, $x_2^{(n-1)}(a_1) = y^{(n-1)}(a_1,m_2)$ can be continued to $t = a_2$ and has

$$B < m < m_2 \le x_2^{(p)}(a_2).$$

Likewise, if $m_1 < m < B$ and is sufficiently large and negative in the sense of Lemma 13.3 so that $y(t,m_1)$ is the unique solution of (13.2), (13.3) then from Remark 13.1 the solution $x_1(t)$ of (9.1) satisfying the initial conditions $x_1^{(i)}(a_1) = 0$, $0 \le i \le n-2$, $x_1^{(n-1)}(a_1) = y^{(n-1)}(a_1,m_1)$ exists as far as $t = a_2$ and has

$$x_1^{(p)}(a_2) \le m_1 < m < B.$$

Let $x(t)$ be any solution of (9.1) satisfying $x^{(i)}(a_1) = 0$, $0 \le i \le n-2$ and $x_1^{(n-1)}(a_1) < x^{(n-1)}(a_1) < x_2^{(n-1)}(a_1)$, then from the uniqueness of the BVP it follows that

$$x_1^{(p)}(t) < x^{(p)}(t) < x_2^{(p)}(t), \quad a_1 < t < a_2$$

and hence

$$x_1^{(i)}(t) < x^{(i)}(t) < x_2^{(i)}(t); \quad a_1 < t < a_2, \ 0 \le i \le p.$$

This also implies that $x(t)$ exists as far as $t = a_2$. Now, by standard theorems on continuity with respect to initial conditions and the intermediate value theorem it follows that there exists a solution of (9.1), (2.7).

**Theorem 13.6**    In the conclusions (2) and (3) of Theorem 13.2 "at most one" can be replaced by "has a unique".

**Proof.**   We need to develop results parallel to Lemmas 13.3, 13.4 and Remark 13.1, then the proof is similar to that of Theorem 13.5.

**Remark 13.2**    The inequalities $(a_2-a_1) < t_p(\bar{K})$, $(a_2-a_1) < z_p^*$ and $(a_2-a_1) < z_p^{**}$ in the above results are obviously the best possible.

**Corollary 13.7**    Suppose that the function $f(t,x_0,x_1,\ldots,x_q)$ for all $(t,x_0,x_1,\ldots,x_q) \in [a_1,a_2] \times R^{q+1}$ is

(1)   nondecreasing in $x_i$, $0 \leq i \leq q$ then the BVP (9.1), (2.7) has a unique solution

(2)   nondecreasing in $x_i$ for i odd and nonincreasing in $x_i$ for i even $0 \leq i \leq q$, then the BVP (9.1), (2.8) for n odd has a unique solution

(3)   nonincreasing in $x_i$ for i odd and nondecreasing in $x_i$ for i even $0 \leq i \leq q$, then the BVP (9.1), (2.8) for n even has a unique solution.

**Remark 13.3**    The BVP $x'' = (x + \frac{1}{4} x^2)$, $x(-1) = x(1) = 1$ has exactly two solutions thus in Corollary 13.7 the increasing nature cannot be replaced on the compact subset of $[a_1,a_2] \times R^{q+1}$.

## COMMENTS AND BIBLIOGRAPHY

The results of this section are taken from [1] and cover some particular cases discussed in [2,3,5,6]. Several other interesting results when q is not necessarily less than p are given in [4].

164

1.  Agarwal, R. P. "Some new results on two-point problems for higher order differential equations", Funkcialaj Ekvacioj, to appear.
2.  Agarwal, R. P. "Best possible length estimates for nonlinear boundary value problems", Bull. Inst. Math. Acad. Sinica 9, 235-248 (1981).
3.  Bailey, P. B., Shampine, L. F. and Waltman, P. E. "The first and second boundary value problems for nonlinear second order equations", J. Diff. Equs. 2, 399-411 (1966).
4.  Erbe, L. "Boundary value problems for ordinary differential equations", Rocky Mountain J. Math. 4, 709-729 (1971).
5.  Peterson, A. C. "Existence-uniqueness for two-point boundary value problems for nth order nonlinear differential equations", Rocky Mountain J. Math. 7, 103-109 (1977).
6.  Schrader, K. and Umamaheswaram, S. "Existence theorems for higher order boundary value problems", Proc. Amer. Math. Soc. 47, 89-97 (1975).

## 14. MONOTONE CONVERGENCE AND FURTHER EXISTENCE

In [3], a fixed point theorem for isotone operators in partially ordered spaces has been proved which is distinguished for its constructive character. Its slightly strengthened version is given by the following :

Theorem 14.1  Let $(E, \leq)$ be a partially ordered space and $x_0 \leq y_0$ be two elements of E. $[x_0, y_0]$ denotes the interval $\{x \in E : x_0 \leq x \leq y_0\}$. Let $T : [x_0, y_0] \to E$ be isotone $(T(x) \leq T(y))$, whenever $x \leq y$) and let it possess the properties

(i)   $x_0 \leq T(x_0)$

(ii)   the (nondecreasing) sequence $\{T^m(x_0)\}$ where $T^0(x_0) = x_0$, $T^{m+1}(x_0) = T[T^m(x_0)]$ for each $m = 0,1,2,\ldots$ is well defined, i.e., $T^m(x_0) \leq y_0$ for each natural m

(iii)   the sequence $\{T^m(x_0)\}$ has sup $x \in E$, i.e., $T^m(x_0) \uparrow x$

(iv)   $T^{m+1}(x_0) \uparrow T(x)$.

(i)'   $T(y_0) \leq y_0$

(ii)'   the (nonincreasing) sequence $\{T^m(y_0)\}$ is well defined, i.e., $T^m(y_0) \geq x_0$ for each natural m

(iii)'   the sequence $\{T^m(y_0)\}$ has inf $y \in E$, i.e., $T^m(y_0) \downarrow y$

(iv)'   $T^{m+1}(y_0) \downarrow T(y)$.

Then, $x = T(x)$ and for any other fixed point z of T, $x \leq z$ is true.  (Then, $y = T(y)$ and for any other fixed point z of T, $z \leq y$ is valid.)

Moreover, if T possesses both properties (i) and (i)', then the sequences $\{T^m(x_0)\}$, $\{T^m(y_0)\}$ are well defined and if, further, T has the properties (iii), (iii)' and (iv), (iv)'  then

$$x_0 \leq T(x_0) \leq \cdots \leq T^m(x_0) \leq \cdots \leq x \leq y \leq \cdots \leq T^m(y_0)$$
$$\leq \cdots \leq T(y_0) \leq y_0$$

and $x = T(x)$, $y = T(y)$, also any other fixed point $z \in [x_0, y_0]$ of $T$ satisfies $x \leq z \leq y$.

Though this theorem is very simple, it can successfully be used to obtain some constructive existence theorems. The proof of the Viswanatham lemma [11] and the proof of an existence theorem by Bange [2] have served to form the basis of this theorem. Here we shall use Theorem 14.1 to prove few existence results for the equation (9.1) together with some of the boundary conditions (2.3)-(2.9). For this, we need

Lemma 14.2   Let $M > 0$ and $\{x_m(t)\}$ be a sequence of functions in $C^{(n)}[a_1, a_r]$ such that $|x_m(t)| \leq M$ and $|x_m^{(n)}(t)| \leq M$ for all $m$. Then, there exists a subsequence $\{x_{m(j)}(t)\}$ such that $\{x_{m(j)}^{(i)}(t)\}$ converges uniformly on $[a_1, a_r]$ for each $i$, $0 \leq i \leq n-1$.

Definition 14.1   We call a function $\phi \in C^{(n)}[a_1, a_r]$ a __lower solution__ of (9.1) provided

(14.1)     $\phi^{(n)}(t) \geq f(t, \phi(t), \phi'(t), \ldots, \phi^{(q)}(t))$, $t \in [a_1, a_r]$.

Similarly, a function $\psi \in C^{(n)}[a_1, a_r]$ is called an __upper solution__ of (9.1) if

(14.2)     $\psi^{(n)}(t) \leq f(t, \psi(t), \psi'(t), \ldots, \psi^{(q)}(t))$, $t \in [a_1, a_r]$.

In the space $C[a_1, a_r]$ we shall consider the maximum norm and introduce a partial ordering as follows : For $x, y \in C[a_1, a_r]$ we say that $x \leq y$ if and only if $(-1)^{k_{i+1} + \cdots + k_r + (r-i)}[y(t) - x(t)] \leq 0$ for all t e $[a_i, a_{i+1}]$, $1 \le i \le r-1$. Thus, from Lemma 3.2 the set of indices $1,\ldots,r-1$ can be divided into two groups $H_1 = \{i \ e \ \{1, \ldots, r-1\} : k_{i+1} + \ldots + k_r + (r-i)$ is odd$\}$, $H_2 = \{i \ e \ \{1, \ldots, r-1\} : k_{i+1} + \ldots + k_r + (r-i)$ is even$\}$ such that if $i \ e \ H_1(H_2)$, then $g_2(t,s) \le 0$, $a_i \le t \le a_{i+1}$ and $y(t) \ge x(t)$ on the same interval $(g_2(t,s) \ge 0$, $y(t) \le x(t)$ for $a_i \le t \le a_{i+1})$.

**Theorem 14.3** With respect to the BVP (9.1), (2.4) we assume that

    (i)    $q = 0$ and the function $f(t,x)$ is nonincreasing in x for each t e $[a_i, a_{i+1}]$, i e $H_1$ and nondecreasing in x for each t e $[a_i, a_{i+1}]$, i e $H_2$

    (ii)   there exist lower and upper solutions $x_0(t)$, $y_0(t)$ of (9.1) such that

(14.3)
$$x_0 \le y_0$$

and

(14.4)
$$P_{n-1,x_0} \le P_{n-1} \le P_{n-1,y_0}$$

where $P_{n-1,x_0}(t)$ and $P_{n-1,y_0}(t)$ are the polynomials of degree $(n-1)$, satisfying

$$P_{n-1,x_0}^{(j)}(a_i) = x_0^{(j)}(a_i)$$

and

$$P_{n-1,y_0}^{(j)}(a_i) = y_0^{(j)}(a_i); \quad 1 \le i \le r, \ 0 \le j \le k_i$$

respectively.

Then, the sequences $\{x_m\}$, $\{y_m\}$ where $x_m(t)$ and $y_m(t)$ are defined by the iterative schemes

$$x_{m+1}(t) = P_{n-1}(t) + \int_{a_1}^{a_r} g_2(t,s)f(s,x_m(s))ds$$

$$y_{m+1}(t) = P_{n-1}(t) + \int_{a_1}^{a_r} g_2(t,s)f(s,y_m(s))ds; \quad m = 0,1,\ldots$$

are well defined and $\{x_m\}$ converges to an element $x \in C[a_1,a_r]$, $\{y_m\}$ converges to an element $y \in C[a_1,a_r]$ (the convergence being in the norm of $C[a_1,a_r]$). Further

$$x_0 \leq x_1 \leq \cdots \leq x_m \leq \cdots \leq x \leq y \leq \cdots \leq y_m \leq \cdots \leq y_1 \leq y_0,$$

$x(t)$ and $y(t)$ are solutions of the BVP (9.1), (2.4) and each solution $z(t)$ of this problem which is such that $z \in [x_0, y_0]$ satisfies $x \leq z \leq y$.

Proof.    First, we shall show that the operator $T : C[a_1,a_r] \to C^{(n)}[a_1,a_r]$ defined in (9.2) is isotone. For this, let $x,y \in C[a_1,a_r]$ and $x \leq y$, then $x(t) \leq y(t)$ for $t \in [a_i, a_{i+1}]$, $i \in H_1$ and

$$T(x)(t) = P_{n-1}(t) + \int_{a_1}^{a_r} g_2(t,s)f(s,x(s))ds$$

$$\leq P_{n-1}(t) + \int_{a_1}^{a_r} g_2(t,s)f(s,y(s))ds$$

$$= T(y)(t).$$

This  together with the reversed inequality in the interval $[a_i, a_{i+1}]$ for $i \in H_2$ gives $T(x) \leq T(y)$. Hence, $T$ is isotone.

Next, since $x_0(t)$ is a lower solution, for $t \in [a_i, a_{i+1}]$, $i \in H_1$ we have

$$x_0(t) = P_{n-1,x_0}(t) + \int_{a_1}^{a_r} g_2(t,s)x_0^{(n)}(s)ds$$

$$\leq P_{n-1}(t) + \int_{a_1}^{a_r} g_2(t,s)f(s,x_0(s))ds$$

$$= T(x_0)(t).$$

This together with the reversed inequality in the interval $[a_i, a_{i+1}]$ for $i \in H_2$ implies that $x_0 \leq T(x_0)$ in $C[a_1, a_r]$. The inequality $T(y_0) \leq y_0$ can be proved analogously. Thus, the conditions (i), (i)' of Theorem 14.1 hold and in conclusion the sequences $\{T^m(x_0)\}$, $\{T^m(y_0)\}$ are well defined.

Since $T^m(x_0) = T(T^{m-1}(x_0))$, we have $T^m(x_0) = x_m$ and $T^m(y_0) = y_m$. The sequence $\{x_m(t)\}$ is nondecreasing and bounded from above by $y_0(t)$ in $[a_i, a_{i+1}]$, $i \in H_1$ and is nonincreasing and bounded from below by $y_0(t)$ in $[a_i, a_{i+1}]$, $i \in H_2$. A similar argument holds for the sequence $\{y_m(t)\}$. Hence, the sequences $\{x_m(t)\}$, $\{y_m(t)\}$ are uniformly bounded on $[a_1, a_r]$.

Now on using the above monotonicity properties, it is easy to verify that

$$y_0^{(n)}(t) \leq f(t, y_0(t)) \leq x_{m+1}^{(n)}(t) \leq f(t, x_0(t)) \leq x_0^{(n)}(t), \quad t \in [a_1, a_r]$$

for all $m$. A similar argument holds for the sequence $\{y_m^{(n)}(t)\}$. Hence, the sequences $\{x_m^{(n)}(t)\}$, $\{y_m^{(n)}(t)\}$ are also uniformly bounded on $[a_1, a_r]$.

Thus, from Lemma 14.2 there exist subsequences $\{x_{m(j)}(t)\}$, $\{y_{m(j)}(t)\}$ which converge uniformly on $[a_1, a_r]$. However, since the sequences $\{x_m(t)\}$, $\{y_m(t)\}$ are monotonic, we conclude that the whole sequences $\{x_m(t)\}$, $\{y_m(t)\}$ converge uniformly on $[a_1, a_r]$ to some $x(t)$, $y(t)$ such that $x, y \in C[a_1, a_r]$, i.e., $T^m(x_0) \uparrow x$ and $T^m(y_0) \downarrow y$.

Finally, from the continuity of $T$ it is obvious that $T^{m+1}(x_0) = T(T^m(x_0)) \uparrow T(x)$ and $T^{m+1}(y_0) = T(T^m(y_0)) \downarrow T(y)$.

Hence, the conditions of Theorem 14.1 are satisfied and the conclusions of Theorem 14.3 follow.

While Theorem 14.3 bears a constructive character, the next
result brings a pure existence statement.

**Theorem 14.4**  Suppose that q = 0 and the following hold

   (i)   condition (ii) of Theorem 14.3

   (ii)       $y_0^{(n)}(t) \leq f(t,x) \leq x_0^{(n)}(t)$

           for all $(t,x) \in [a_1, a_r] \times W$, where

       $W = \{x : x_0(t) \leq x \leq y_0(t)\} \cup \{x : y_0(t) \leq x \leq x_0(t)\}$.

Then, the BVP (9.1), (2.4) has a solution x(t) such that x lies in the
interval $[x_0, y_0]$.

**Proof.**   Consider the operator $T : C[a_1, a_r] \to C[a_1, a_r]$ defined in (9.2).
The interval $[x_0, y_0]$ is a closed, convex and bounded subset of $C[a_1, a_r]$.
When $x_0 \leq x \leq y_0$, then $(t, x(t)) \in [a_1, a_r] \times W$ and for all $t \in [a_i, a_{i+1}]$,
$i \in H_1$ we have

$$T(x)(t) = P_{n-1}(t) + \int_{a_1}^{a_r} g_2(t,s) f(s, x(s)) ds$$

$$\leq P_{n-1, y_0}(t) + \int_{a_1}^{a_r} g_2(t,s) y_0^{(n)}(s) ds$$

$$= y_0(t).$$

Similarly, the following inequalities can be proved

$$T(x)(t) \geq y_0(t); \quad t \in [a_i, a_{i+1}], \quad i \in H_2$$

$$T(x)(t) \geq x_0(t); \quad t \in [a_i, a_{i+1}], \quad i \in H_1$$

$$T(x)(t) \leq x_0(t); \quad t \in [a_i, a_{i+1}], \quad i \in H_2.$$

Thus, we find that $T([x_0, y_0]) \subseteq [x_0, y_0]$.  By the Schauder fixed point theorem there is a fixed point of T in $[x_0, y_0]$ and that completes the proof of Theorem 14.4.

Theorem 14.5   With respect to the BVP (7.1), (7.7), (7.8) we assume that the constants $\alpha_0$, $\alpha_1$, $\beta_0$, $\beta_1$ are nonnegative and

    (i)   the function f is independent of x' and nonincreasing in x for each t $\epsilon$ $[a_1, a_2]$

    (ii)  there exist lower and upper solutions $x_0(t)$, $y_0(t)$ of (7.1), (7.7), (7.8) satisfying $x_0(t) \leq y_0(t)$ on $[a_1, a_2]$.

Then, the sequences $\{x_m(t)\}$, $\{y_m(t)\}$ defined by the iterative schemes

$$x_{m+1}(t) = \ell(t) + \int_{a_1}^{a_r} g_1(t,s)f(s,x_m(s))ds$$

$$y_{m+1}(t) = \ell(t) + \int_{a_1}^{a_r} g_1(t,s)f(s,y_m(s))ds; \quad m = 0,1,\ldots$$

converge to solutions $x(t)$, $y(t)$ of the BVP (7.1), (7.7), (7.8). Further

$$x_0(t) \leq x_1(t) \leq \cdots \leq x_m(t) \leq \cdots \leq x(t) \leq y(t) \leq \cdots \leq y_m(t) \leq \cdots \leq y_1(t) \leq y_0(t)$$

and each solution z(t) of this problem which is such that $x_0(t) \leq z(t) \leq y_0(t)$ satisfies $x(t) \leq z(t) \leq y(t)$.

Proof.   Let the space $C[a_1, a_2]$ be as in Theorem 14.3 with r = 2, then $H_1 = \{1\}$, $H_2 = \emptyset$. Thus, if x,y $\epsilon$ $C[a_1, a_2]$ and $x \leq y$, then from Lemma 3.1, $g_1(t,s) \leq 0$, $a_1 \leq t \leq a_2$ and $x(t) \leq y(t)$ on the same interval.

Next, since $x_0(t)$ is a lower solution of (7.1), (7.7), (7.8) it is easy to verify that $\ell_{x_0}(t) \leq \ell(t)$, where $\ell_{x_0}(t)$ is the straight line satisfying

$$\alpha_0 \ell_{x_0}(a_1) - \alpha_1 \ell'_{x_0}(a_1) = \alpha_0 x_0(a_1) - \alpha_1 x_0'(a_1)$$

$$\beta_0 \ell_{x_0}(a_2) + \beta_1 \ell'_{x_0}(a_2) = \beta_0 x_0(a_2) + \beta_1 x'_0(a_2).$$

Similarly, the straight line $\ell_{y_0}(t)$ is defined and it is easy to see that $\ell_{y_0}(t) \geq \ell(t)$.

Now as in Theorem 14.3 it is easy to check that the operator $T : C[a_1,a_2] \rightarrow C^{(2)}[a_1,a_2]$ defined in (7.10) satisfies the conditions of Theorem 14.1.

Our next result is for the BVP (9.1), (2.6). For this, we need to consider the following four cases

    (i)   n is even, k is odd

    (ii)  n is even, k is even

    (iii) n is odd, k is odd

    (iv) n is odd, k is even.

We shall discuss only the case (i), since results for the other three cases can be stated analogously.

In the space $C^{(q)}[a_1,a_2]$ we shall consider the norm $\|x\| = \max\limits_{0 \leq i \leq q} \{ \max\limits_{a_1 \leq t \leq a_2} |x^{(i)}(t)| \}$ and introduce a partial ordering as follows : For $x,y \in C^{(q)}[a_1,a_2]$ we say that $x \leq y$ if and only if $x^{(i)}(t) \leq y^{(i)}(t)$, $0 \leq i \leq k$ or $k < i$ (odd) $\leq q$; $y^{(i)}(t) \leq x^{(i)}(t)$, $k < i$ (even) $\leq q$ for all $t \in [a_1,a_2]$. Thus, for the case (i), from Lemma 3.5, $g^{(i)}(t,s) \leq 0$ if $0 \leq i \leq k$ or $k < i$ (odd) $\leq q$ and $g^{(i)}(t,s) \geq 0$ if $k < i$ (even) $\leq q$ for all $a_1 \leq s$, $t \leq a_2$.

Theorem 14.6   With respect to the BVP (9.1), (2.6) we assume that

    (i)   n is even, k is odd and the function $f(t,x_0,x_1,\ldots,x_q)$ is nonincreasing in $x_i$ for all $0 \leq i \leq k$, $k < i$ (odd) $\leq q$ and nondecreasing in $x_i$ for all $k < i$ (even) $\leq q$

    (ii)  there exist lower and upper solutions $x_0(t)$, $y_0(t)$ of (9.1) such that

$$x_0 \le y_0$$

and

$$P_{n-1,x_0} \le P_{n-1} \le P_{n-1,y_0}$$

where $P_{n-1,x_0}(t)$ and $P_{n-1,y_0}(t)$ are the polynomials of degree (n-1), satisfying

$$P_{n-1,x_0}^{(i)}(a_1) = x_0^{(i)}(a_1), \ 0 \le i \le k-1; \ P_{n-1,x_0}^{(i)}(a_2) = x_0^{(i)}(a_2), \ k \le i \le n-1$$

and

$$P_{n-1,y_0}^{(i)}(a_1) = y_0^{(i)}(a_1), \ 0 \le i \le k-1; \ P_{n-1,y_0}^{(i)}(a_2) = y_0^{(i)}(a_2), \ k \le i \le n-1$$

respectively.

Then, the sequences $\{x_m\}$, $\{y_m\}$ where $x_m(t)$ and $y_m(t)$ are defined by the iterative schemes

$$x_{m+1}(t) = P_{n-1}(t) + \int_{a_1}^{a_2} g_4(t,s)f(s,\underline{x}_m(s))ds$$

$$y_{m+1}(t) = P_{n-1}(t) + \int_{a_1}^{a_2} g_4(t,s)f(s,\underline{y}_m(s))ds; \ m = 0,1,\ldots$$

are well defined and $\{x_m\}$ converges to an element $x \in C^{(q)}[a_1,a_2]$, $\{y_m\}$ converges to an element $y \in C^{(q)}[a_1,a_2]$ (the convergence being in the norm of $C^{(q)}[a_1,a_2]$). Further

$$x_0 \le x_1 \le \cdots \le x_m \le \cdots \le x \le y \le \cdots \le y_m \le \cdots \le y_1 \le y_0,$$

$x(t)$ and $y(t)$ are solutions of the BVP (9.1), (2.6) and each solution $z(t)$ of this problem which is such that $z \in [x_0,y_0]$ satisfies $x \le z \le y$.

Proof. As in Theorem 14.3 it is easy to verify that the operator $T : C^{(q)}[a_1,a_2] \to C^{(n)}[a_1,a_2]$ defined by

$$T(x)(t) = P_{n-1}(t) + \int_{a_1}^{a_2} g_4(t,s)f(s,\underline{x}(s))ds$$

satisfies the conditions of Theorem 14.1.

**Theorem 14.7**   Suppose that n is even, k is odd and the following hold

   (i)   condition (ii) of Theorem 14.6

   (ii)   $y_0^{(n)}(t) \leq f(t,x_0,x_1,\ldots,x_q) \leq x_0^{(n)}(t)$

   for all $(t,x_0,x_1,\ldots,x_q) \in [a_1,a_2] \times W$, where

   $W = \{(x_0,x_1,\ldots,x_q) : x_0^{(i)}(t) \leq x_i \leq y_0^{(i)}(t), 0 \leq i \leq q\} \cup$

   $\{(x_0,x_1,\ldots,x_q) : y_0^{(i)}(t) \leq x_i \leq x_0^{(i)}(t), 0 \leq i \leq q\}.$

Then, the BVP (9.1), (2.6) has a solution $x(t)$ such that x lies in the interval $[x_0,y_0]$.

**Proof.**   The proof is similar to that of Theorem 14.4.

Finally, we remark that in view of Lemmas 3.6 and 3.7 similar results for the BVPs (9.1), (2.7) and (9.1), (2.8) when $q \leq p$ can be stated without much difficulty.

**Example 14.1**   For the BVP

(14.5)        $x'' = -x^{2p+1} + t$   (p, a positive integer)

(14.6)        $x(0) = 0, \quad x'(1) = 1$

with $x_0(t) = t(t-1)$, $y_0(t) = t$ it is easy to verify that the conditions of Theorem 14.5 are satisfied.

For this BVP, we have

$$g_1(t,s) = \begin{cases} -s & s \leq t \\ -t & t \leq s \end{cases}$$

and $\ell(t) = t$. Thus, the iterative schemes

$$x_{m+1}(t) = t - \int_0^t s(s - x_m^{2p+1}(s))ds - t\int_t^1 (s - x_m^{2p+1}(s))ds$$

$$y_{m+1}(t) = t - \int_0^t s(s - y_m^{2p+1}(s))ds - t\int_t^1 (s - y_m^{2p+1}(s))ds; \quad m = 0,1,\ldots$$

converge monotonically to the solutions $x(t)$, $y(t)$ of the BVP (14.5), (14.6). Moreover, $t(t-1) \leq x(t) \leq y(t) \leq t$.

## COMMENTS AND BIBLIOGRAPHY

Theorems 14.3 and 14.4 are modelled after Šeda [10] and generalize some of the results obtained in [6,9]. Theorem 14.5 is adapted from [8]. For several other related results see [1,4,5,7]. In conclusion the knowledge of Sgn $g^{(i)}(t,s)$; $0 \leq i \leq q$, $a \leq s$, $t \leq b$ plays an important role in obtaining monotonic convergence.

1.  Agarwal, R. P. and Krishnamoorthy, P. R. "Boundary value problems for nth order ordinary differential equations", Bull. Inst. Math. Acad. Sinica 7, 211-230 (1979).
2.  Bange, D. W. "Periodic solutions of a quasilinear parabolic differential equation", J. Diff. Equs. 17, 61-72 (1975).
3.  Edwards, R. E. Functional Analysis - Theory and Applications, Holt, Rinehart and Winston, New York, 1965.
4.  Eisenfeld, J. and Lakshmikantham, V. "On a boundary value problem for a class of differential equations with a deviating argument", J. Math. Anal. Appl. 51, 158-164 (1975).
5.  Jagdish Chandra, "A comparison result for a boundary value problem for a class of nonlinear differential equations with a deviating argument", J. Math. Anal. Appl. 47, 573-577 (1974).
6.  Klaasen, G. A. "Differential inequalities and existence theorems for second and third order boundary value problems", J. Diff.

Equs. <u>10</u>, 529-537 (1971).

7.  Krishnamoorthy, P. R. and Agarwal, R. P. "Higher order boundary value problems for differential equations with deviating arguments", Math. Seminar Notes <u>7</u>, 253-260 (1979).

8.  Schmitt, K. "A nonlinear boundary value problem", J. Diff. Equs. <u>9</u>, 527-537 (1970).

9.  Schmitt, K. "Boundary value problems and comparison theorems for ordinary differential equations", SIAM J. Appl. Math. <u>26</u>, 670-678 (1974).

10. Šeda, V. "Two remarks on boundary value problems for ordinary differential equations", J. Diff. Equs. <u>26</u>, 278-290 (1977).

11. Viswanatham, B. "A generalization of Bellman's lemma", Proc. Amer. Math. Soc. <u>14</u>, 15-18 (1963).

## 15. UNIQUENESS IMPLIES EXISTENCE

In Section 2, we have noticed that the uniqueness of solutions of the linear BVP (2.1), (2.2) implies the existence of solutions. The argument employed in proving this assertion is algebraic and is based on the linear structure of the fundamental system of solutions of (2.12) and the linearity of the boundary conditions (2.2). The question of whether or not nonlinear equation (9.1) could have this property will be discussed here. For this, we need Kamke's convergence theorem [2].

**Theorem 15.1**   Assume that in the equations

$$(15.1)_m \qquad\qquad x^{(n)} = f_m(t,\underline{x}); \ m = 0,1,\ldots$$

the functions $f_m(t,\underline{x})$ are continuous on $J \times R^{q+1}$, where $J$ is an interval and assume that $\lim_{m \to \infty} f_m(t,\underline{x}) = f_0(t,\underline{x})$ uniformly on each compact subset of $J \times R^{q+1}$. Assume that the sequence $\{t_m\} \subset J$ with $\lim_{m \to \infty} t_m = t_0$ and that, for each integer $m \geq 1$, $x_m(t)$ is a solution of $(15.1)_m$ which is defined on a maximum interval $J_m \subset J$ with $t_m \ e \ J_m$. Further, assume that $\lim_{m \to \infty} x_m^{(i)}(t_m) = x_i$ for each $0 \leq i \leq n-1$. Then, there is a subsequence $\{x_{m(j)}(t)\}$ of $\{x_m(t)\}$ and there is a solution $x_0(t)$ of $(15.1)_0$ defined on a maximal interval $J_0 \subset J$ such that $t_0 \ e \ J_0$, $x_0^{(i)}(t_0) = x_i$ for each $0 \leq i \leq n-1$, and such that for any compact interval $[c,d] \subset J_0$ it follows that $[c,d] \subset J_{m(j)}$ for all sufficiently large $m(j)$ and $\lim x_{m(j)}^{(i)}(t) = x_0^{(i)}(t)$ uniformly on $[c,d]$ for each $0 \leq i \leq n-1$.

**Corollary 15.2**   Assume that for the equation (9.1) the following conditions are satisfied

(A)   $f(t,\underline{x})$ is continuous on $J \times R^{q+1}$

(B)   all solutions of (9.1) extend to $J$

(C) solutions of initial value problems for (9.1) are unique.

Then, solutions of (9.1) depend continuously on initial conditions in the sense that, given any solution $x(t)$ of (9.1), given any $t_0 \in J$, given any compact interval $[c,d] \subset J$, and given any $\varepsilon > 0$, there is a $\delta > 0$ such that, if $y(t)$ is any solution of (9.1) with $|y^{(i)}(t_0) - x^{(i)}(t_0)| < \delta$ for $0 \le i \le n-1$, then $|y^{(i)}(t) - x^{(i)}(t)| < \varepsilon$ on $[c,d]$ for $0 \le i \le n-1$.

Lasota and Opial [13] in 1967 published the first result establishing that uniqueness implies existence of solutions of BVPs for nonlinear differential equations. Their result is contained in the following :

**Theorem 15.3**   Assume that the differential equation (7.1) satisfies the conditions (A), (B) and (C) of Corollary 15.2 on $[a,b) \times R^2$. In addition assume that the following uniqueness condition for two point BVPs is satisfied :

> (D)  For any $a \le a_1 < a_2 < b$ and any solutions $x(t)$ and $y(t)$ of (7.1), $x(a_i) = y(a_i)$ for $i = 1,2$ implies $x(t) \equiv y(t)$.

Then, for any $a \le a_1 < a_2 < b$ and any $A, B \in R$ the BVP (7.1), (7.2) has a solution.

**Proof.**   Let $a \le a_1 < a_2 < b$ and $A, B \in R$ be given and let $x(t;p)$ be the solution of (7.1) satisfying the initial conditions $x(a_1) = A$, $x'(a_1) = p$. Then, to show that (7.1), (7.2) has a solution it suffices to show that $S \equiv \{x(a_2;p) : -\infty < p < \infty\}$ is the whole real line. From Corollary 15.2 it follows that $x(a_2;p)$ is a continuous function of $p$ which in turn implies that $S$ is an interval. Consequently, to show that $S = R$ it suffices to show that $S$ is neither bounded above nor below. Assume that $S$ is bounded above and that $\beta \in R$ is such that $x(a_2;p) < \beta$ for all $p \in R$. Let $y(t)$ be the solution of (7.1) satisfying the initial conditions $y(a_2) = \beta$, $y'(a_2) = 0$. Since $\beta \notin S$, $y(a_1) \ne A$. Assume first that $y(a_1) > A$ and let $\{x_m(t)\}$ be the sequence of solutions of (7.1) defined by $x_m(t) = x(t;m)$. Then, it follows from the uniqueness hypothesis (D) that $x_1(t) \leq x_m(t) \leq y(t)$ on $[a_1, a_2]$ for all $m \geq 1$. Thus, if $M > 0$ is such that $|x_1(t)| \leq M$ and $|y(t)| \leq M$ on $[a_1, a_2]$, then $|x_m(t)| \leq M$ on $[a_1, a_2]$ and for each $m \geq 1$ there is a $t_m$ e $(a_1, a_2)$ such that $(a_2 - a_1)|x_m'(t_m)| = |x_m(a_2) - x_m(a_1)| \leq 2M$. Consequently, we may select a subsequence, relabelled as the original sequence, such that $\{t_m\}$, $\{x_m(t_m)\}$ and $\{x_m'(t_m)\}$ all converge. Then, by Theorem 15.1 and hypothesis (B) there is a further subsequence such that $\{x_m^{(i)}(t)\}$ converges uniformly on $[a_1, a_2]$ for $i = 0,1$. However, this is impossible since $x_m'(a_1) = x'(a_1;m) = m$. We conclude that it is not possible to have $y(a_1) > A$ and therefore $y(a_1) < A$. In this case, if $\{x_m(t)\}$ is again the sequence with $x_m(t) = x(t;m)$ for each $m$ and if $a_2 < a_3 < b$, then $x_1(t) \leq x_m(t) \leq y(t)$ on $[a_2, a_3]$. This leads to the same contradiction as before which now forces us to the conclusion that S cannot be bounded above. A similar argument leads to the conclusion that S is not bounded below. Thus, $S = R$ and the BVP (7.1), (7.2) has a solution.

Remark 15.1    The proof of Theorem 15.3 makes use of the availability of an interval on each side of $t = a_2$ and in fact the result is false when stated in terms of a closed interval $[a,b]$. In order to prove this, let the interval be $[0,\pi]$. We note that the implicit equation

$$\phi + \frac{1}{2} p \tan^{-1}\phi = q, \quad p > -2$$

has a unique solution $\phi(p,q)$. Further, it is easy to verify that the family of all solutions of the differential equation

(15.2)    $$x'' = -x + \frac{1}{2} \tan^{-1}\phi(\sin t, \ x \sin t + x'\cos t)$$

can be represented by

$$x(t) = \alpha \cos t + \beta \sin t + \frac{1}{2} \tan^{-1}\beta$$

where $\alpha$ and $\beta$ are arbitrary constants. Thus, the boundary conditions (7.2) lead to the following system of equations

$$\alpha \cos a_1 + \beta \sin a_1 + \frac{1}{2} \tan^{-1}\beta = A$$

$$\alpha \cos a_2 + \beta \sin a_2 + \frac{1}{2} \tan^{-1}\beta = B.$$

Eliminating $\alpha$, we obtain

(15.3)  $\beta \sin(a_2 - a_1) + \frac{1}{2}(\cos a_1 - \cos a_2)\tan^{-1}\beta = B \cos a_1 - A \cos a_2$.

Now, if $0 \le a_1 < a_2 < \pi$ or $0 < a_1 < a_2 \le \pi$, then obviously

$$\sin(a_2 - a_1) > 0, \quad \cos a_1 - \cos a_2 > 0$$

and hence for every pair A, B of real numbers the BVP (15.2), (7.2) has a unique solution. However, when $a_1 = 0$ and $a_2 = \pi$, then the equation (15.3) reduces to

$$\tan^{-1}\beta = B + A.$$

As before, this assures the uniqueness of solutions of the BVP (15.2), (7.2) but at the same time it proves that they do not exist if $|B + A| > \frac{\pi}{2}$ .

We turn now to the question of extending Theorem 15.3 to n point BVP (9.1), (2.3). For this, we need the Brouwer theorem on the invariance of domain [18].

Theorem 15.4   If U is an open subset of $R^n$, n dimensional Euclidean space and $\phi : U \to R^n$ is one to one and continuous on U, then $\phi$ is a homeomorphism and $\phi(U)$ is an open subset of $R^n$.

Theorem 15.5   Assume that the differential equation (9.1) satisfies the conditions (A), (B) and (C) of Corollary 15.2 on (a,b) × $R^{q+1}$.   In addition assume that the following uniqueness condition for n point BVPs is satisfied :

(D)   For any $a < a_1 < \ldots < a_n < b$ and any solutions x(t) and $y(t)$ of (9.1), $x(a_i) = y(a_i)$ for $1 \leq i \leq n$ implies $x(t) \equiv y(t)$ on $(a,b)$.

Then, given any $a < a_1 < \ldots < a_n < b$ and any solution $x(t)$ of (9.1) there is a $\varepsilon > 0$ such that $|t_i - a_i| < \varepsilon$ and $|x(a_i) - x_i| < \varepsilon$ for $1 \leq i \leq n$ implies that there is a solution $y(t)$ of (9.1) with $y(t_i) = x_i$ for $1 \leq i \leq n$. Furthermore, if for each $i$, $1 \leq i \leq n$, $\{t_{i(j)}\} \subset (a,b)$ is a sequence with $\lim_{j \to \infty} t_{i(j)} = a_i$ and if $\{y_j(t)\}$ is a sequence of solutions of (9.1) such that $\lim_{j \to \infty} y_j(t_{i(j)}) = x(a_i)$ for each $1 \leq i \leq n$, then $\lim_{j \to \infty} y_j^{(k)}(t) = x^{(k)}(t)$ uniformly on compact subintervals of $(a,b)$ for each $0 \leq k \leq n-1$.

**Proof.** Let $\Delta = \{(a_1,\ldots,a_n) : a < a_1 < \ldots < a_n < b\}$. Then, $\Delta \times R^n$ is an open subset of $R^{2n}$. Let $a_0$ be an arbitrary but fixed point in $(a,b)$ and define $\phi : \Delta \times R^n \to R^{2n}$ by

(15.4) $\qquad \phi(a_1,\ldots,a_n,c_1,\ldots,c_n) = (a_1,\ldots,a_n,x(a_1),\ldots,x(a_n))$

where $x(t)$ is the unique solution of (9.1) satisfying the initial conditions $x^{(i-1)}(a_0) = c_i$, $1 \leq i \leq n$. Then, it follows from Corollary 15.2 that $\phi$ is continuous on $\Delta \times R^n$. Assume that $(a_1,\ldots,a_n,c_1,\ldots,c_n)$ and $(t_1,\ldots,t_n,d_1,\ldots,d_n)$ in $\Delta \times R^n$ are such that

$$\phi(a_1,\ldots,a_n,c_1,\ldots,c_n) = \phi(t_1,\ldots,t_n,d_1,\ldots,d_n).$$

It follows from the definition of $\phi$ in (15.4) that $a_i = t_i$ for $1 \leq i \leq n$ and that $x(a_i) = y(a_i)$ for $1 \leq i \leq n$ where $x(t)$ and $y(t)$ are the solutions of (9.1) with $x^{(i-1)}(a_0) = c_i$ and $y^{(i-1)}(a_0) = d_i$ for $1 \leq i \leq n$. However, by hypothesis (D) the condition $x(a_i) = y(a_i)$ for $1 \leq i \leq n$ implies $x(t) \equiv y(t)$ on $(a,b)$. Thus, we also have $c_i = d_i$ for $1 \leq i \leq n$ and $\phi$ is one to one on $\Delta \times R^n$. It follows from Theorem 15.4 that $\phi(\Delta \times R^n)$

is an open set in $R^{2n}$ and that $\phi^{-1}$ is continuous on $\phi(\Delta \times R^n)$. The first assertion of the theorem follows from the fact that $\phi(\Delta \times R^n)$ is open. The second assertion, that is the continuity of solutions with respect to boundary values, follows from the continuity of $\phi^{-1}$ and the continuity of solutions of (9.1) with respect to initial values as asserted in Corollary 15.2.

**Corollary 15.6**    Assume that the differential equation (9.1) satisfies the conditions (A), (B), (C) and (D) of Theorem 15.5. Then, if $a < a_1 < \ldots < a_{n-1} < b$ and if $x(t)$ and $y(t)$ are distinct solutions of (9.1) such that $x(a_i) = y(a_i)$ for $1 \le i \le n-1$, it follows that $x(t) - y(t)$ changes sign at each $a_i$, $1 \le i \le n-1$.

**Proof.**    Since $x(t)$ and $y(t)$ are assumed to be distinct solutions, $x(t) - y(t)$ is zero only at the points $a_i$, $1 \le i \le n-1$. Assume that for some $j$, $1 \le j \le n-1$, $x(t) - y(t)$ has the same sign on each side of $a_j$, say, $x(t) - y(t) > 0$ on the right and left open intervals adjacent to $a_j$. Choose an $a_0 \in (a,b)$ with $a_0 \ne a_i$ for $1 \le i \le n-1$. Then, it follows from Theorem 15.5 that for $\varepsilon > 0$ sufficiently small (9.1) has a solution $x(t;\varepsilon)$ with $x(a_i;\varepsilon) = x(a_i)$ for $0 \le i \le n-1$, $i \ne j$ and $x(a_j;\varepsilon) = x(a_j) - \varepsilon$. Furthermore, $\lim_{\varepsilon \to 0} x(t;\varepsilon) = x(t)$ uniformly on compact subintervals of $(a,b)$. Thus, if $\delta > 0$ is chosen so that $[a_j - \delta, a_j + \delta] \subset (a,b)$ and $a_i \notin [a_j - \delta, a_j + \delta]$ for $i \ne j$, then for sufficiently small $\varepsilon > 0$ we will have $x(a_j \pm \delta;\varepsilon) > y(a_j \pm \delta)$ and $x(a_j;\varepsilon) = x(a_j) - \varepsilon$. Hence, $x(t;\varepsilon) - y(t)$ will have distinct zeros at $t = a_i$; $1 \le i \le n-1$, $i \ne j$ and two zeros in $(a_j - \delta, a_j + \delta)$. This contradicts the hypothesis (D) and we conclude that $x(t) - y(t)$ must change sign at $a_j$.

We are now ready to prove the result which is an extension of Theorem 15.3 to differential equations of arbitrary orders.

**Theorem 15.7**    Assume that the differential equation (9.1) satisfies the conditions (A), (B), (C) and (D) of Theorem 15.5 and the additional hypothesis :

> (E)  If $[c,d]$ is a compact interval of $(a,b)$ and $\{x_m(t)\}$ is a
> sequence of solutions of (9.1) such that $|x_m(t)| \le M$ on $[c,d]$
> for some $M > 0$ and all $m = 1,2,\ldots$, then there is a sub-
> sequence $\{x_{m(j)}(t)\}$ such that $\{x_{m(j)}^{(i)}(t)\}$ converges uniformly
> on $[c,d]$ for each $0 \le i \le n-1$.

Then, for any $a < a_1 < \ldots < a_n < b$ and any real numbers $A_i$, $1 \le i \le n$
the BVP (9.1), (2.3) has a solution.

Proof.  Let a particular BVP (9.1), (2.3) be specified.  Let $x_0(t)$ be
an arbitrary but fixed solution of (9.1) on $(a,b)$.  It will suffice to
show that there is a solution $x_1(t)$ such that $x_1(a_1) = A_1$ and $x_1(a_j) = x_0(a_j)$ for $2 \le j \le n$.  For, if this has been done, we can repeat the
argument starting with $x_1(t)$ to obtain a solution $x_2(t)$ such that
$x_2(a_2) = A_2$ and $x_2(a_j) = x_1(a_j)$ for $1 \le j \le n$, $j \ne 2$, and proceeding
in this way, obtain a solution of (9.1), (2.3) in n steps.

For this, from Theorem 15.5 we note that the set

(15.5)    $S = \{x(a_1) : x(t)$ is a solution of (9.1) and
$$x(a_j) = x_0(a_j), \ 2 \le j \le n\}$$

is an open subset of the reals.  If it can be shown that S is also
closed, it will follow that $S = R$ and that there is a solution $x_1(t)$ of
(9.1) with $x_1(a_1) = A_1$ and $x_1(a_j) = x_0(a_j)$, $2 \le j \le n$.  Hence, assume
$r_0 \ \epsilon \ R$ is a limit point of S which is not contained in S and that
$\{r_m\} \subset S$ is such that $\lim r_m = r_0$.  We can assume that $\{r_m\}$ is strictly
monotone and to deal with a specific case assume the sequence is strictly
increasing.  Let $\{x_m(t)\}$ be the sequence of solutions of (9.1) such that
$x_m(a_1) = r_m$ and $x_m(a_j) = x_0(a_j)$ for $2 \le j \le n$.  It follows from Corollary
15.6 that $\{x_m(t)\}$ is strictly increasing on $(a,a_2)$, is strictly decreasing on $(a_2,a_3)$, and is alternately strictly increasing and decreasing on

the successive intervals $(a_3, a_4), \ldots, (a_n, b)$.

If the sequence $\{x_m(t)\}$ is bounded on any compact interval of $(a,b)$ it follows from hypothesis (E) and Theorem 15.1 that there is a solution $y(t)$ of (9.1) and a subsequence of $\{x_m(t)\}$ which converges uniformly to $y(t)$ on each compact subinterval of $(a,b)$. However, this would imply that $y(a_1) = r_0$ and $y(a_j) = x_0(a_j)$ for $2 \leq j \leq n$ which contradicts the assumption that $r_0 \notin S$. Thus, the sequence $\{x_m(t)\}$ is not bounded on any compact subinterval of $(a,b)$. Let $z(t)$ be a solution of (9.1) with $z(a_1) = r_0$, for example, let $z(t)$ be the solution of (9.1) satisfying the initial conditions $z(a_1) = r_0$, $z^{(i)}(a_1) = 0$, $1 \leq i \leq n-1$. Using the monotoneity properties of $\{x_m(t)\}$ observed above and the fact that the sequence is unbounded on each of the intervals $(a, a_1), (a_1, a_2), \ldots, (a_{n-1}, a_n)$ and $(a_n, b)$ we conclude that for sufficiently large $m$, $x_m(t) - z(t)$ has a zero in both a right and left neighborhood of $a_1$ and in disjoint neighborhoods of $a_2, \ldots, a_n$. By hypothesis (D), this implies $x_m(t) \equiv z(t)$ for all sufficiently large $m$. From this contradiction we conclude that $S$ is closed and the proof of the theorem is complete.

Remark 15.2    In [3] Hartman has proved that if (9.1) satisfies the conditions (A), (B), (C) and (D) and if for any $a < a_1 < \ldots < a_n < b$ and any real numbers $A_i$, $1 \leq i \leq n$ the BVP (9.1), (2.3) has a solution, then for any $a < a_1 < \ldots < a_r < b$ and any real numbers $A_{j+1,i}$; $1 \leq i \leq r$, $0 \leq j \leq k_i$ the BVP (9.1), (2.4) has a unique solution. Thus, from Theorem 15.7 we have the following :

Theorem 15.8    Assume that the differential equation (9.1) satisfies the conditions (A)-(E) of Theorem 15.7. Then, each $r$ point BVP, i.e., for any $a < a_1 < \ldots < a_r < b$ and any real numbers $A_{j+1,i}$; $1 \leq i \leq r$, $0 \leq j \leq k_i$ the BVP (9.1), (2.4), has a unique solution.

Remark 15.3    With some simple adjustment of the arguments Theorem 15.7 as well as Theorem 15.8 remains valid if the conditions (A)-(E) are stated on $[a,b) \times R^{q+1}$ or $(a,b] \times R^{q+1}$.

Remark 15.4    The proposition "in Theorem 15.7 conditions (A)-(D) only
are sufficient" is proved in Theorem 15.3 for n = 2, and on using different arguments for n = 3 it is given in [8]; however for n > 3 it remains
an open problem, except when q = 0.  For this, let $[c,d] \subset (a,b)$ and
$\{x_m(t)\}$ be a sequence of solutions of (9.1) such that $|x_m(t)| \leq M$ on
[c,d] for some M > 0 and all m = 1,2,... .  Then, from the continuity
of f(t,x) it follows that there exists a constant C such that

$$|x_m^{(n)}(t)| = \max_{c \leq t \leq d} |f(t,x_m(t))| \leq C$$

for all m = 1,2,... .  Thus, from Lemma 14.2 there exists a subsequence
$\{x_{m(j)}(t)\}$ such that $\{x_{m(j)}^{(i)}(t)\}$ converges uniformly on [c,d] for each
$0 \leq i \leq$ n-1.

Remark 15.5    A number of generalizations of Theorems 15.3, 15.5 and
15.7 have been given both in terms of generalizing the boundary conditions and in terms of weakening the uniqueness hypothesis for solutions
of initial value problems as well as of BVPs.  For example, with more
involved arguments Theorem 15.3 can be proved without the assumption
that solutions of initial value problems are unique, i.e., condition
(C) can be dropped [14,19].  Generalizations to more general boundary
conditions than (7.2) or weakening the uniqueness hypothesis (D) in
Theorem 15.3 have been considered in [1,12,15,17].  Results similar to
Theorem 15.3 for the two dimensional vector systems have been obtained
in [20].  Klaasen [11] has proved a version of Theorem 15.5 in which
condition (C) is not assumed.  In [4] Theorem 15.7 has been proved for
vector systems u' = F(t,u), u = $(x_0, x_1, \ldots, x_{n-1})$, F = $(f_0, f_1, \ldots, f_{n-1})$
which are such that in a solution vector u(t) = $(x_0(t), x_1(t), \ldots, x_{n-1}(t))$
the functions $x_0(t), x_1(t), \ldots, x_{n-1}(t)$ are successive pseudoderivatives
of $x_0(t)$.  The vector form of equation (9.1) is a special case of such
a system.  If (9.1) satisfies hypotheses (A)-(D) of Theorem 15.5 and if
n point BVPs are locally solvable, i.e., if (9.1) also satisfies the
hypothesis :

    $(E_1)$ For each $t_0 \in$ (a,b) there is an open interval $J(t_0) \subset$ (a,b)

containing $t_0$ such that on $J(t_0)$ all n point BVPs have
solutions

then, all n point BVPs for (9.1) on (a,b) have solutions, i.e., the
conclusion of Theorem 15.7 holds. Obviously, by Theorem 15.7 if (9.1)
satisfies (A)-(E), then $(E_1)$ is satisfied. Conversely, it can be shown
that if (9.1) satisfies (A)-(D) and $(E_1)$, then (E) is satisfied. In
[4] these results are proved for systems u' = F(t,u). Klaasen [10] has
proved Theorem 15.7 by methods quite different from those used in [4].
In [6] the hypothesis (D) of Theorem 15.7 has been replaced by

($D_1$) all two point BVPs for (9.1) on (a,b) have solutions, and all
(n-1) point BVPs for (9.1) on (a,b) have at most one solution.

In an another paper Šeda [16] has proved Theorem 15.7 for n = 4, q = 3
in which condition (C) is dropped and (E) is replaced by the hypothesis :

($E_2$) There exists a K > 0 such that

$$f(t,x,x',x'',x''') \geq 0 (f(t,x,x',x'',x''') \leq 0)$$

for all $(t,x,x') \in (a,b) \times R^2$ and $x'' \geq K$, $x''' \geq K$
($x'' \leq -K$, $x''' \leq -K$).

Recently in [5] a result similar to that of Theorem 15.8 for the
right $(m_1,...,m_r)$ focal point BVP has been obtained, where the right
$(m_1,...,m_r)$ focal point BVP is defined as follows :

Definition 15.1   Let $2 \leq r \leq n$ and $m_i$, $1 \leq i \leq r$ be positive integers
such that $\sum_{i=1}^{r} m_i = n$. Let $s_0 = 0$ and for $1 \leq j \leq r$, $s_j = \sum_{i=1}^{j} m_i$. A
BVP for (9.1) with boundary conditions of the form

(15.6)         $$x^{(i)}(a_j) = A_{ij} ; \quad s_{j-1} \leq i \leq s_j - 1, \quad 1 \leq j \leq r$$

where $a < a_1 < ... < a_r < b$ is called a right $(m_1,...,m_r)$ focal point
BVP for (9.1) on (a,b).

Thus, in particular the boundary conditions (2.5) and (2.6) are right $(1,\ldots,1)$ and $(k,n-k)$ focal point boundary conditions.

**Theorem 15.9** Assume that the differential equation (9.1) satisfies the conditions (A), (B), (C) and (E) of Theorem 15.7 and the additional hypothesis :

$(D_2)$ Each right $(1,\ldots,1)$ focal point BVP for (9.1) on $(a,b)$ has at most one solution.

Then, each right $(m_1,\ldots,m_r)$ focal point BVP for (9.1) on $(a,b)$ has a unique solution.

## COMMENTS AND BIBLIOGRAPHY

Theorems 15.3, 15.5 and 15.7 are taken from [9]. Remark 15.1 is due to Lasota [12]. For another example proving the Remark 15.1 see [7].

1.  Chow, Shui-Nee and Lasota, A. "On boundary value problems for ordinary differential equations", J. Diff. Equs. 14, 326-327 (1973).
2.  Hartman, P. Ordinary Differential Equations, Wiley, New York, 1964.
3.  Hartman, P. Unrestricted n-parameter families", Rend. Circ. Mat. Palermo (2) 7, 123-142 (1958).
4.  Hartman, P. "On n-parameter families and interpolation problems for nonlinear ordinary differential equations", Trans. Amer. Math. Soc. 154, 201-226 (1971).
5.  Henderson, J. "Existence of solutions of right focal point boundary value problems for ordinary differential equations", Nonlinear Analysis : Theory, Methods and Appl. 5, 989-1002 (1981).
6.  Henderson, J. and Jackson, L. "Existence and uniqueness of k-point boundary value problems for ordinary differential equations", J. Diff. Equs. 48, 373-385 (1983).
7.  Jackson, L. K. "Subfunctions and second-order ordinary differential inequalities", Advances in Math. 2, 307-363 (1968).
8.  Jackson, L. K. and Schrader, K. W. "Existence and uniqueness of solutions of boundary value problems for third order differential equations", J. Diff. Equs. 9, 46-54 (1971).

9.  Jackson, L. K. "Boundary value problems for ordinary differential equations", Studies in Mathematics, The Mathematical Association of America 1977, 93-125.

10. Klaasen, G. "Existence theorems for boundary value problems for nth order ordinary differential equations", Rocky Mountain J. Math. 3, 457-472 (1973).

11. Klaasen, G. "Continuous dependence for N-point boundary value problems", SIAM J. Appl. Math. 29, 99-102 (1975).

12. Lasota, A. "Boundary value problems for second order differential equations", Seminar on Differential Equations and Dynamic Systems II, Lecture Notes in Mathematics 144, Springer-Verlag, New York, 1970.

13. Lasota, A. and Opial, Z. "On the existence and uniqueness of solutions of a boundary value problem for an ordinary second order differential equation", Colloq. Math. 18, 1-5 (1967).

14. Schrader, K. "Existence theorems for second order boundary value problems", J. Diff. Equs. 5, 572-584 (1969).

15. Schrader, K. and Waltman, P. "An existence theorem for nonlinear boundary value problems", Proc. Amer. Math. Soc. 21, 653-656 (1969).

16. Šeda, V. "On a boundary value problem of the fourth order", Časopis pro pěstováni matematiky, roč. 106, 65-74 (1981).

17. Shampine, L. F. "Existence and uniqueness for nonlinear boundary value problems", J. Diff. Equs. 5, 346-351 (1969).

18. Spanier, E. H. Algebraic Topology, McGraw-Hill, New York, 1966.

19. Waltman, P. "Existence and uniqueness of solutions to a nonlinear boundary value problem", J. Math. Mech. 18, 585-586 (1968).

20. Waltman, P. "Existence and uniqueness of solutions of boundary value problems for two-dimensional systems of nonlinear differential equations", Trans. Amer. Math. Soc. 153, 223-234 (1971).

## 16. COMPACTNESS CONDITION AND GENERALIZED SOLUTIONS

The proposition mentioned in Remark 15.4 is that "conditions (A)-(D) of Theorem 15.5 are sufficient to imply the conclusion of Theorem 15.7". This is indeed the case when the differential equation (9.1) is of second or third order. In fact for these particular cases conditions (A)-(D) imply the compactness condition (E). To show this we note that Theorem 15.1 can be stated as follows :

**Theorem 16.1**  Assume that for the differential equation (9.1) the conditions (A), (B) and (C) of Theorem 15.5 are satisfied. Then, if $\{x_m(t)\}$ is a sequence of solutions of (9.1) and $[c,d]$ is a compact subinterval of $(a,b)$, either there is a subsequence $\{x_{m(j)}(t)\}$ such that $\{x_{m(j)}^{(i)}(t)\}$ converges uniformly on $[c,d]$ for each $0 \le i \le n-1$, or

$$\sum_{i=0}^{n-1} |x_m^{(i)}(t)| \to \infty \text{ uniformly on } [c,d] \text{ as } m \to \infty.$$

If the differential equation (9.1) is of second order and satisfies hypotheses (A), (B) and (C) of Theorem 15.5, then (9.1) also satisfies condition (E). For in this case, if $\{x_m(t)\}$ is a sequence of solutions of (9.1) with $|x_m(t)| \le M$ on $[c,d] \subset (a,b)$ for each $m \ge 1$, then for each $m \ge 1$ there is a $t_m \in [c,d]$ such that $|x_m(t_m)| + |x_m'(t_m)| \le M + \frac{2M}{d-c}$. It then follows from Theorem 16.1 that hypothesis (E) is satisfied.

If the differential equation (9.1) is of third order such an immediate appeal to Theorem 16.1 is not possible and we need the following lemmas.

**Lemma 16.2**  Assume that the differential equation (9.1) is of order three and satisfies hypothesis (A) of Theorem 15.5. Then, given any compact interval $[c,d] \subset (a,b)$ and any fixed $M > 0$, there is a $\delta > 0$ such that $[a_1,a_2] \subset [c,d]$, $a_2 - a_1 \le \delta$, and $|\alpha| \le M$ implies that (9.1) has solutions $x_1(t)$ and $x_2(t)$ satisfying $x_i(a_1) = x_i(a_2) = \alpha$ for $i=1,2$,

$x_1'(a_1) = x_2'(a_2) = 0$ and $\left| x_i'(t) \right| \leq 1$, $\left| x_i''(t) \right| \leq 1$ on $[a_1, a_2]$ for $i = 1, 2$.

**Proof.** The proof follows from Theorem 9.1.

**Lemma 16.3** Assume that the differential equation (9.1) is of order three and satisfies hypotheses (A), (B), (C) and (D) of Theorem 15.5. Then, solutions of two point BVPs for (9.1) when they exist are unique.

**Proof.** This is a special case of Theorem 17.4 which will be proved in the next section.

**Lemma 16.4** Let $x(t) \in C^{(2)}[\alpha, \beta]$ and assume $\left| x(t) \right| \leq M$ on $[\alpha, \beta]$. There is an $N > 0$ depending on M and $\beta - \alpha$ such that, if $\max\{\left| x'(t) \right|, \left| x''(t) \right|\} > N$ for all $\alpha \leq t \leq \beta$, then $x'(t_0) = 0$ for some $t_0$ with $\alpha < t_0 < \beta$.

**Proof.** The proof is elementary.

**Theorem 16.5** Assume that $n = 3$ and the differential equation (9.1) satisfies hypotheses (A), (B), (C) and (D) of Theorem 15.5. Then, (9.1) also satisfies condition (E).

**Proof.** We assume that the conclusion of the theorem is false. Then, by Theorem 16.1 there is a compact interval $[c,d] \subset (a,b)$, a $M > 0$ and a sequence of solutions $\{x_m(t)\}$ such that $\left| x_m(t) \right| \leq M$ on $[c,d]$ for all $m \geq 1$ and such that $\sum_{i=0}^{2} \left| x_m^{(i)}(t) \right| \to \infty$ uniformly on $[c,d]$. Let $c \leq a_1 < a_2 < a_3 < a_4 \leq d$ be such that $a_4 - a_1 \leq \delta$, where $\delta$ is as defined in Lemma 16.2. By Lemma 16.4 there is an $N > 0$ such that, if $\max\{\left| x_m'(t) \right|, \left| x_m''(t) \right|\} > N$ for each $t \in [c,d]$, then $x_m'(t)$ has a zero on $(a_1, a_2)$, on $(a_2, a_3)$ and on $(a_3, a_4)$. Furthermore, we can assume that $N > 1$. From the fact that $\left| x_m'(t) \right| + \left| x_m''(t) \right| \to \infty$ uniformly on $[c,d]$ we can conclude that there is a positive integer $m_0$ such that

$$\max\{\left| x_{m_0}'(t) \right|, \left| x_{m_0}''(t) \right|\} > N$$

on $[c,d]$. Let $a_1 < t_1 < a_2 < t_2 < a_3 < t_3 < a_4$ be such that $x'_{m_0}(t_i) = 0$
for $i = 1,2,3$. Then, $\left| x''_{m_0}(t_i) \right| > N > 1$ for $i = 1,2,3$.

There are two cases to consider. First, if $x_{m_0}(t_i) = x_{m_0}(t_j)$ with
$t_i < t_j$, then $x_{m_0}(t)$ is the solution of the differential equation (9.1),
satisfying

$$x(t_i) = x(t_j) = x_{m_0}(t_i), \quad x'(t_i) = 0.$$

Since $t_j - t_i < \delta$, it follows from Lemma 16.2 that $\left| x'_{m_0}(t) \right| \leq 1$ and
$\left| x''_{m_0}(t) \right| \leq 1$ on $[t_i, t_j]$ which is a contradiction to $\left| x''_{m_0}(t_i) \right| > N > 1$.
If $x_{m_0}(t_i) \neq x_{m_0}(t_j)$ for $t_i \neq t_j$, then it suffices to assume

$$x_{m_0}(t_1) < x_{m_0}(t_2) < x_{m_0}(t_3)$$

since the same argument applies to the other possible orderings of the
values of $x_{m_0}(t_i)$; $i = 1,2,3$. If $x''_{m_0}(t_2) > N$, there is a $\tau_1$, $t_1 < \tau_1 < t_2$
such that $x_{m_0}(\tau_1) = x_{m_0}(t_2)$. If $x''_{m_0}(t_2) < -N$, there is a $\tau_2$, $t_2 < \tau_2 < t_3$
such that $x_{m_0}(\tau_2) = x_{m_0}(t_2)$. In either case Lemma 16.2 is again applied
to obtain a contradiction.

It seems very difficult, if not impossible, to extend the method
of Theorem 16.5 to equations of higher orders. In the following result
we shall prove that for equation (9.1) of arbitrary order n the hypotheses (A), (B) and (D) of Theorem 15.5 do imply a weaker type of compactness condition for the solutions of equation (9.1).

Theorem 16.6    Assume that the differential equation (9.1) satisfies
hypotheses (A), (B) and (D) of Theorem 15.5. Then, if $[c,d]$ is a compact
subinterval of $(a,b)$ and $\{x_m(t)\}$ is a sequence of solutions of (9.1)
which is uniformly bounded on $[c,d]$, it follows that the sequence
$\{V_c^d(x_m)\}$ of total variations of the functions $x_m(t)$ on $[c,d]$ is bounded.

Proof. Assume on the contrary that (9.1) satisfies (A), (B) and (D) but that there is a compact interval $[c,d] \subset (a,b)$ and a sequence $\{x_m(t)\}$ of solutions of (9.1) with $|x_m(t)| \leq M$ on $[c,d]$ for all $m \geq 1$ and with $\{V_c^d(x_m)\}$ unbounded. Then, by choosing a subsequence and relabelling if necessary, we can assume $V_c^d(x_m) \to \infty$ as $m \to \infty$.

The condition $V_c^d(x_m) \to \infty$ implies that $\lim\limits_{m \to \infty} \sum\limits_{i=0}^{n-1} |x_m^{(i)}(t)| = \infty$ uniformly on $[c,d]$. For, if it were not the case, then from Theorem 16.1a a subsequence $\{x_{m(j)}(t)\}$ could be chosen such that $\{x_{m(j)}^{(i)}(t)\}$ would converge uniformly on $[c,d]$ for each $i = 0,1,\ldots,n-1$. Obviously, this would contradict $V_c^d(x_m) \to \infty$.

Next, consider the BVP (9.1), (2.3) with $A_i = \alpha$, $1 \leq i \leq n$. From Theorem 9.1 there is a fixed $\delta > 0$ such that for any $\alpha$ with $|\alpha| \leq M$ and any points $c \leq a_1 < \ldots < a_n \leq d$ with $a_n - a_1 \leq \delta$ there is a solution $x(t)$ with $|x(t)| \leq M+1$ on $[a_1,a_n]$ and $|x^{(i)}(t)| \leq 1$ on $[a_1,a_n]$ for $1 \leq i \leq n-1$. Here M is the bound on the sequence $\{x_m(t)\}$ on $[c,d]$.

Let $m_0$ be such that $\sum\limits_{i=0}^{n-1} |x_m^{(i)}(t)| > M+n$ on $[c,d]$ for $m \geq m_0$.

Then, for $m \geq m_0$ the graph of $x = x_m(t)$ can intersect a line $x = \alpha$, $|\alpha| \leq M$ in at most n-1 distinct points in a subinterval of $[c,d]$ of length not exceeding $\delta$ where $\delta > 0$ is as above. For, if $x_m(a_i) = \alpha$ at points $a_i$, $1 \leq i \leq n$ with $c \leq a_1 < \ldots < a_n \leq d$ and $a_n - a_1 \leq \delta$, it would follow from the uniqueness hypothesis (D) that $x_m(t)$ would coincide on $[a_1,a_n]$ with the solution obtained above which would contradict $\sum\limits_{i=0}^{n-1} |x_m^{(i)}(t)| > M+n$ on $[a_1,a_n]$. Thus, if k is the integer such that $(k-1)\delta \leq d - c < k\delta$, then for $m \geq m_0$ the graph of $x = x_m(t)$ cannot intersect a line $x = \alpha$, $|\alpha| \leq M$ in more than (n-1)k distinct points. For $-M \leq \alpha \leq M$, let $h_m(\alpha)$ be the number of distinct points $t \in [c,d]$ such that $x_m(t) = \alpha$. Then, from [1,p.270] it follows that $V_c^d(x_m) = \int_{-M}^{M} h_m(\alpha)d\alpha$

This leads to the contradiction that $V_c^d(x_m) \leq 2M(n-1)k$ for all $m \geq m_0$ and the proof of the theorem is complete.

Corollary 16.7   Assume that the differential equation (9.1) satisfies hypotheses (A), (B) and (D) of Theorem 15.5. Then, if [c,d] is a compact subinterval of (a,b) and if $\{x_m(t)\}$ is a sequence of solutions of (9.1) which is uniformly bounded on [c,d], there is a subsequence $\{x_{m(j)}(t)\}$ which converges pointwise on [c,d] and $y(t) = \lim\limits_{j \to \infty} x_{m(j)}(t)$ is of bounded variation on [c,d].

Proof.   The result follows from the Helly selection theorem [7,p.398] and the fact that $\{x_m(t)\}$ and $\{V_c^d(x_m)\}$ are bounded.

Now to prove the proposition we shall consider another possible approach.   For this, if the equation (9.1) satisfies the hypotheses (A)-(D) of Theorem 15.5, then it is not difficult to show that the compactness condition (E) is equivalent to the following :

   ($E_3$)   If $\{x_m(t)\}$ is a sequence of solutions of (9.1) which is monotone and bounded on some compact interval $[c,d] \subset (a,b)$, then $\lim\limits_{m \to \infty} x_m(t)$ is a solution of (9.1) on [c,d].

Definition 16.1   A function $\phi$ defined on an interval $J \subset (a,b)$ is said to be a <u>generalized solution</u> on J if for each set of points $a_1 < a_2 < \cdots < a_n$ contained in J and any solution x(t) of (9.1), the inequalities $(-1)^{n+i}[x(a_i) - \phi(a_i)] < 0$, $1 \leq i \leq n$ imply $x(t) < \phi(t)$ on $J \cap [a_n,b)$ and $(-1)^{n+1}[x(t) - \phi(t)] < 0$ on $J \cap (a,a_1]$, and the inequalities $(-1)^{n+i} \times [x(a_i) - \phi(a_i)] > 0$, $1 \leq i \leq n$ imply $x(t) > \phi(t)$ on $J \cap [a_n,b)$ and $(-1)^{n+1}[x(t) - \phi(t)] > 0$ on $J \cap (a,a_1]$.

Theorem 16.8   Assume that the differential equation (9.1) satisfies hypotheses (A) - (D) of Theorem 15.5 and $\lim\limits_{m \to \infty} x_m(t) = \phi(t)$ on $J \subset (a,b)$, where $\{x_m(t)\}$ is a sequence of solutions of (9.1).   Then, $\phi(t)$ is a generalized solution of (9.1) on J.

Proof.  Assume that for $a_1 < a_2 < \ldots < a_n$ contained in J there is a
solution $x(t)$ of (9.1) such that $(-1)^{n+i}[x(a_i) - \phi(a_i)] < 0$ for $1 \leq i \leq n$
but that also $x(a_0) > \phi(a_0)$ for some $a_0 > a_n$ in J.  Then, since
$\lim_{m \to \infty} x_m(t) = \phi(t)$, there is a solution $x_m(t)$ of (9.1) such that
$(-1)^{n+i}[x(a_i) - x_m(a_i)] < 0$ for $1 \leq i \leq n$ and $x(a_0) > x_m(a_0)$.  This
contradicts hypothesis (D).  The remaining inequalities can be proved
in a similar way.

Thus, the limit of a bounded monotone sequence of solutions $\{x_m(t)\}$
of (9.1) satisfying (A) - (D) of Theorem 15.5 is a generalized solution.

Lemma 16.9  Assume that the differential equation (9.1) satisfies
hypothesis (A) of Theorem 15.5 and that $\phi \in C^{(n-1)}[c,d]$, where $[c,d]$ is
a compact subinterval of $(a,b)$.  Assume that $M > 0$ is such that
$|\phi^{(i)}(t)| \leq M$ on $[c,d]$ for $0 \leq i \leq n-1$.  Then, there exists a $\delta > 0$ such
that, for any $c \leq a_1 < a_2 < \ldots < a_n \leq d$ with $a_n - a_1 \leq \delta$ , (9.1) has a
solution $x(t)$ with $x(a_i) = \phi(a_i)$ for $1 \leq i \leq n$ and $|x^{(j)}(t)| \leq 2M$ on
$[a_1, a_n]$ for $0 \leq j \leq n-1$.  Furthermore, $\delta$ can be chosen  in such a way
that, for each fixed set $a_1 < a_2 < \ldots < a_n$ satisfying the above condi-
tions, there is a $\varepsilon > 0$ such that for any $x_i$, $1 \leq i \leq n$ with $|x_i - \phi(a_i)|$
$< \varepsilon$, $1 \leq i \leq n$, (9.1) has a solution $x(t)$ satisfying $x(a_i) = x_i$, $1 \leq i \leq n$
and $|x^{(j)}(t)| \leq 3M$ on $[a_1, a_n]$ for $0 \leq j \leq n-1$.

Proof.  The proof follows from Corollary 9.8.

Theorem 16.10  Assume that the differential equation (9.1) satisfies
hypotheses (A) and (D) of Theorem 15.5 and that $\lim_{m \to \infty} x_m(t) = \phi(t)$ on
$[c,d] \subset (a,b)$, where $\{x_m(t)\}$ is a sequence of solutions of (9.1).  Then,
if $\phi \in C^{(n-1)}[c,d]$, $\phi(t)$ is a solution of (9.1) on $[c,d]$ and $\lim_{m \to \infty} x_m^{(j)}(t) =$
$\phi^{(j)}(t)$ uniformly on $[c,d]$ for each $0 \leq j \leq n-1$.

**Proof.** Let $M > 0$ be such that $|\phi^{(j)}(t)| \leq M$ on $[c,d]$ for $0 \leq j \leq n-1$. By Lemma 16.9 there is a $\delta > 0$ such that, if $c \leq a_1 < a_2 < \ldots < a_n \leq d$ is a fixed set of points with $a_n - a_1 \leq \delta$, there is a $\varepsilon > 0$ with the property that $|x_i - \phi(a_i)| < \varepsilon$ for $1 \leq i \leq n$ implies that (9.1) has a solution $x(t)$ satisfying $x(a_i) = x_i$ for $1 \leq i \leq n$ and $|x^{(j)}(t)| \leq 3M$ on $[a_1, a_n]$ for $0 \leq j \leq n-1$. It follows that there is an $N > 0$ such that $m \geq N$ implies $|x_m(a_i) - \phi(a_i)| < \varepsilon$ for $1 \leq i \leq n$. Hence, by hypothesis (D) and the choice of $\varepsilon$, $|x_m^{(j)}(t)| \leq 3M$ on $[a_1, a_n]$ for $0 \leq j \leq n-1$ and $m \geq N$. From this the conclusion follows.

Let $\phi(t)$ be a real valued function defined on $(c,d)$. At a point $t_0 \in (c,d)$ where $\phi(t)$ has a finite right hand limit $\phi(t_0+0)$, we define

$$D^1\phi(t_0+0) = \lim_{t \to t_0^+} \frac{\phi(t) - \phi(t_0+0)}{t - t_0}$$

provided the limit exists. The left derivative $D^1\phi(t_0-0)$ is similarly defined. Likewise, if $\phi(t_0+0)$ and $D^1\phi(t_0+0)$ exist and are finite

$$D^2\phi(t_0+0) = \lim_{t \to t_0^+} \left\{ \frac{2}{(t-t_0)^2} [\phi(t) - \phi(t_0+0) - D^1\phi(t_0+0)(t-t_0)] \right\}$$

provided the limit exists. In general, if the limits defining $\phi(t_0+0)$ and $D^j\phi(t_0+0)$, $1 \leq j \leq k-1$ exist and are finite, we define

$$D^k\phi(t_0+0) = \lim_{t \to t_0^+} \left\{ \frac{k!}{(t-t_0)^k} [\phi(t) - \phi(t_0+0) - \sum_{j=1}^{k-1} \frac{D^j\phi(t_0+0)(t-t_0)^j}{j!}] \right\}$$

provided the limit exists. The left derivatives $D^j\phi(t_0-0)$ are defined in a corresponding way.

**Theorem 16.11**  Assume that the differential equation (9.1) satisfies hypotheses (A) and (D) of Theorem 15.5 and that $\phi(t)$ is a bounded generalized solution of (9.1) on $(c,d) \subset (a,b)$. Then, $\phi(t)$ has right and left limits at each point of $(c,d)$ and $D^1\phi(t_0-0)$ and $D^1\phi(t_0+0)$ exist in the extended reals for all $t_0 \in (c,d)$. Furthermore, if at a point $t_0 \in (c,d)$, $D^j\phi(t_0+0)$ exists and is finite for each $1 \le j \le k-1 \le n-2$, then the limit defining $D^k\phi(t_0+0)$ exists in the extended reals. The same assertion applies to the left derivative $D^k\phi(t_0-0)$.

Proof. Assume that for some $t_0 \in (c,d)$, $\lim\inf\limits_{t\to t_0^+} \phi(t) < \lim\sup\limits_{t\to t_0^+} \phi(t)$ and choose a real number $r$ such that $\lim\inf\limits_{t\to t_0^+} \phi(t) < r < \lim\sup\limits_{t\to t_0^+} \phi(t)$.

Then, there exist sequences $\{s_m\}$ and $\{t_m\}$ in $(c,d)$ such that $\lim s_m = \lim t_m = t_0$, $t_0 < s_{m+1} < t_m < s_m$ for each $m \ge 1$, $\lim \phi(s_m) = \lim\sup\limits_{t\to t_0^+} \phi(t)$, and $\lim \phi(t_m) = \lim\inf\limits_{t\to t_0^+} \phi(t)$. Let $x(t)$ be a solution of (9.1) satisfying the initial conditions $x(t_0) = r$ and $x^{(j)}(t_0) = 0$ for $1 \le j \le n-1$. This solution exists on $[t_0, t_0+\delta]$ for some $\delta > 0$, and since $\lim\limits_{t\to t_0} x(t) = r$, there is an $N$ such that $m \ge N$ implies $t_0 < s_m < t_0 + \delta$ and $\phi(s_m) > x(s_m)$, $\phi(t_m) < x(t_m)$. This contradicts the fact that $\phi(t)$ is a generalized solution on $(c,d)$. The existence of $\phi(t_0-0)$ can be proved similarly.

Now assume that for some $t_0 \in (c,d)$ the limit defining $D^1\phi(t_0+0)$ does not exist in the extended reals. Then, choose the real number $r$ such that

$$\lim\inf\limits_{t\to t_0^+} \frac{\phi(t) - \phi(t_0+0)}{t - t_0} < r < \lim\sup\limits_{t\to t_0^+} \frac{\phi(t) - \phi(t_0+0)}{t - t_0} .$$

If $x(t)$ is a solution of (9.1) satisfying the initial conditions $x(t_0) = \phi(t_0+0)$, $x'(t_0) = r$, and $x^{(j)}(t_0) = 0$ for $2 \le j \le n-1$, again sequences $\{s_m\}$ and $\{t_m\}$ can be chosen so that $\lim s_m = \lim t_m = t_0$, $t_0 < s_{m+1} < t_m < s_m$ for each $m \ge 1$, and $\phi(s_m) > x(s_m)$, $\phi(t_m) < x(t_m)$ for all sufficiently large m.  This again contradicts $\phi(t)$ being a generalized solution.  Thus, $D^1\phi(t_0+0)$ and $D^1\phi(t_0-0)$ exist in the extended reals for all $t_0 \in (c,d)$.

Finally, if we assume that for some $t_0 \in (c,d)$, $D^j\phi(t_0+0)$ exists and is finite for each $1 \leq j \leq k-1 \leq n-2$, then by considering a solution of (9.1) satisfying the initial conditions $x(t_0) = \phi(t_0+0)$, $x^{(j)}(t_0) = D^j\phi(t_0+0)$ for $1 \leq j \leq k-1$, $x^{(k)}(t_0) = r$, and $x^{(j)}(t_0) = 0$ for $k+1 \leq j \leq n-1$, we can as above prove that the limit defining $D^k\phi(t_0+0)$ exists in the extended reals.

Corollary 16.12    Assume that the differential equation (9.1) satisfies hypotheses (A) and (D) of Theorem 15.5 and that $\phi(t)$ is a bounded generalized solution of (9.1) on $(c,d) \subset (a,b)$.  Then, $\phi(t)$ has a finite derivative $\phi'(t)$ almost everywhere on $(c,d)$.

Theorem 16.13    Assume that the differential equation (9.1) satisfies hypotheses (A) -(D) of Theorem 15.5.  Let $\{x_m(t)\}$ be a sequence of solutions of (9.1) on $(c,d) \subset (a,b)$ such that $\{x_m(t)\}$ is uniformly bounded on $(c,d)$ and $\lim x_m(t) = \phi(t)$ on $(c,d)$.  Then, if for some $t_0 \in (c,d)$ the derivatives $D^j\phi(t_0+0)$, $1 \leq j \leq n-1$ all exist and are finite or the derivatives $D^j\phi(t_0-0)$, $1 \leq j \leq n-1$ all exist and are finite, it follows that there is a subsequence $\{x_{m(j)}(t)\}$ such that $\{x_{m(j)}^{(i)}(t)\}$ converges uniformly on each compact subinterval of $(a,b)$ for each $0 \leq i \leq n-1$.

Proof.    Assume that for some $t_0 \in (c,d)$ the derivatives $D^j\phi(t_0+0)$, $1 \leq j \leq n-1$ exist and are finite.  Let $p(t)$ be the polynomial

$$p(t) = \phi(t_0+0) + \sum_{j=1}^{n-1} \frac{D^j\phi(t_0+0)(t-t_0)^j}{j!}$$

then, it follows from the definition of $D^{n-1}\phi(t_0+0)$ that given any $\varepsilon > 0$

there is a $\delta > 0$ such that $t_0 + \delta < d$ and

$$|p(t) - \phi(t)| < \frac{\varepsilon(t - t_0)^{n-1}}{(n-1)!}$$

for $t_0 < t \leq t_0 + \delta$. Let $d_0$ be a fixed number satisfying $t_0 < d_0 < d$. By Lemma 16.9 there is a $\delta_0 > 0$ such that for $t_0 < t_1 < t_2 < \ldots < t_n \leq d_0$ with $t_i - t_{i-1} = \eta \leq \delta_0$ for each $1 \leq i \leq n$, (9.1) has a solution $x(t)$ with $x(t_i) = p(t_i)$ for $1 \leq i \leq n$ and $|x^{(j)}(t)| \leq 2M$ on $[t_1, t_n]$ for $0 \leq j \leq n-1$ where $|p^{(j)}(t)| \leq M$ on $[t_0, d_0]$ for $0 \leq j \leq n-1$. Furthermore, there is a $\varepsilon_0 > 0$ such that, if $|x_i - p(t_i)| < \varepsilon_0$ for $1 \leq i \leq n$, then (9.1) has a solution $x(t)$ with $x(t_i) = x_i$ for $1 \leq i \leq n$ and $|x^{(j)}(t)| \leq 3M$ on $[t_1, t_n]$ for $0 \leq j \leq n-1$. It is not difficult to show that with equal spacing $\eta$ between the $t_i$'s a suitable $\varepsilon_0$ has the form $\varepsilon_0 = Mh_n \eta^{n-1}$ where $h_n$ is a fixed constant depending on n. Now as noted above, if we choose $\varepsilon = \frac{Mh_n}{2n^{n-1}}$, there is a $\eta$, $0 < \eta \leq \delta_0$ such that $t_0 < t < t_0 + n\eta$ implies

$$|p(t) - \phi(t)| < \frac{\varepsilon(t-t_0)^{n-1}}{(n-1)!} \leq \frac{\varepsilon_0}{2(n-1)!} \leq \frac{\varepsilon_0}{2} \;.$$

For such a choice of $\eta > 0$, we have $|p(t_i) - \phi(t_i)| \leq \frac{\varepsilon_0}{2}$ for $1 \leq i \leq n$ where $t_i - t_{i-1} = \eta$ for $1 \leq i \leq n$. Consequently, if $N > 0$ is such that $m \geq N$ implies $|x_m(t_i) - \phi(t_i)| < \frac{\varepsilon_0}{2}$ for $1 \leq i \leq n$, then $|p(t_i) - x_m(t_i)| < \varepsilon_0$ for $m \geq N$ and $1 \leq i \leq n$. It follows from our construction and hypothesis (D) that $|x_m^{(j)}(t)| \leq 3M$ on $[t_1, t_n]$ for $0 \leq j \leq n-1$ and all $m \geq N$. The conclusion of the theorem now follows.

Thus we see that, in order to prove that conditions (A) - (D) of Theorem 15.5 imply the compactness condition (E), it is sufficient to prove that, if $\phi(t)$ is the pointwise limit of a bounded sequence of solutions of (9.1) on $(c,d) \subset (a,b)$, then there is at least one $t_0 \in (c,d)$ at which either $D^j \phi(t_0+0)$, $1 \leq j \leq n-1$ or $D^j \phi(t_0-0)$, $1 \leq j \leq n-1$ are finite.

## COMMENTS AND BIBLIOGRAPHY

Theorem 16.5 is taken from [2]. Theorem 16.6 is adapted from Jackson [4]. Schrader [6] has proved that, if $\{x_m(t)\}$ is a uniformly bounded sequence of functions on a compact interval $[c,d]$ and if the functions $x_m(t)$ satisfy only the uniqueness hypothesis (D) of Theorem 15.5 on $[c,d]$, then there is a subsequence which converges pointwise on $[c,d]$. Rest of the results are due to Jackson [3] and Klaasen [5].

1. Hewitt, E. and Stromberg, K. Real and Abstract Analysis. A Modern Treatment of the Theory of Functions of a Real Variables, Springer-Verlag, New York, 1965.
2. Jackson, L. and Schrader, K. "Existence and uniqueness of solutions of boundary value problems for third order differential equations", J. Diff. Equs. 9, 46-54 (1971).
3. Jackson, L. K. "Uniqueness and existence of solutions of boundary value problems for ordinary differential equations", Proc. N.R.L.-M.R.C. Conference, Academic Press, New York, 137-149, 1972.
4. Jackson, L. K. "A compactness condition for solutions of ordinary differential equations", Proc. Amer. Math. Soc. 57, 89-92 (1976).
5. Klaasen, G. "Existence theorems for boundary value problems for nth order ordinary differential equations", Rocky Mountain J. Math. 3, 457-472 (1973).
6. Schrader, K. "A generalization of the Helly selection theorem", Bull. Amer. Math. Soc. 78, 415-419 (1972).
7. Taylor, A. E. General Theory of Functions and Integration, Blaisdell Pub. Comp., Waltham, 1965.

# 17. UNIQUENESS IMPLIES UNIQUENESS

Let the BVP (7.1), (7.2) have two solutions $x_1(t)$ and $x_2(t)$. Since $x_1(a_i) - x_2(a_i) = 0$ for $i = 1,2$ there exists some $t_0 \in (a_1,a_2)$ at which $x_1'(t_0) - x_2'(t_0) = 0$. Thus, $x_1(t)$ and $x_2(t)$ are both solutions of the two BVPs : (7.1) together with

$$(17.1) \qquad\qquad x(a_1) = A, \quad x'(t_0) = m$$

and (7.1) together with

$$(17.2) \qquad\qquad x'(t_0) = m, \quad x(a_2) = B$$

where $m = x_1'(t_0) = x_2'(t_0)$. Thus, if $a_1 < t_1 < a_2$ and uniqueness holds for all BVPs (7.1), (17.1) whenever $t_0 \in (a_1,t_1]$, and if uniqueness holds for all BVPs (7.1), (17.2) then uniqueness holds for all BVPs (7.1), (7.2) also.

The uniqueness of some BVPs implies the uniqueness of different BVPs is particularly important in uniqueness implies existence type of results. For example, from the above observation in Theorem 15.3 the hypothesis (D) can be replaced by

(D') for $a_1 < t_1 < a_2$ the uniqueness holds for all BVPs (7.1), (17.1) whenever $t_0 \in (a_1,t_1]$, and the uniqueness holds for all BVPs (7.1), (17.2) whenever $t_0 \in [t_1,a_2)$.

For the differential equation (9.1) we shall prove some results so that the uniqueness of $m$ point BVPs on $(a,b)$ implies the uniqueness of $r(\leq m)$ point BVPs on $(a,b)$. For this, we need

Theorem 17.1  Let the differential equation (9.1) be such that the conditions (A), (B) and (C) of Theorem 15.5 are satisfied. Further, let $x(t)$, $y(t)$ be solutions of (9.1) with $x(t) - y(t)$ having a zero of order

p, $2 \leq p \leq n-1$ at $t_0 \in (a,b)$, and assume that $x^{(p)}(t_0) > y^{(p)}(t_0)$.
Then, given a $\delta_1$ with $a < t_0 - \delta_1 < t_0 + \delta_1 < b$ there is a $\delta$, $0 < \delta < \delta_1$
and a solution $z(t)$ of (9.1) such that $z(t) - x(t)$ and $z(t) - y(t)$ each
have p distinct zeros on $(t_0-\delta, t_0+\delta)$ and $y(t_0+\delta) < z(t_0+\delta) < x(t_0+\delta)$.
Furthermore, if $[c,d] \subset (a,b)$ and $\varepsilon > 0$ are given, $z(t)$ can be chosen
so that $|z(t) - x(t)| < \varepsilon$ on $[c,d]$ or $|z(t) - y(t)| < \varepsilon$ on $[c,d]$.

Proof.    We will consider only the case in which p is an even integer
since the argument is essentially the same for p odd.

Let m be such that $y^{(p)}(t_0) < m < x^{(p)}(t_0)$ and let $z_1(t)$ be the
solution of the initial value problem for (9.1) satisfying the initial
conditions

$$z_1^{(i)}(t_0) = x^{(i)}(t_0); \ 0 \leq i \leq n-1, \ i \neq p$$

$$z_1^{(p)}(t_0) = m.$$

Then, there is a $0 < \delta < \delta_1$ such that $y(t) < z_1(t) < x(t)$ on
$[t_0-\delta, t_0+\delta] - \{t_0\}$. By Corollary 15.2 there is a $\varepsilon_1 > 0$ such that
if $z_2(t)$ is the solution of (9.1) with initial conditions

$$z_2^{(i)}(t_0) = z_1^{(i)}(t_0); \ 0 \leq i \leq n-1, \ i \neq p-1$$

$$z_2^{(p-1)}(t_0) = z_1^{(p-1)}(t_0) + \varepsilon_1,$$

then

$$y(t_0 \pm \delta) < z_2(t_0 \pm \delta) < x(t_0 \pm \delta).$$

Furthermore, since p-1 is odd and

$$z_1^{(p-1)}(t_0) = x^{(p-1)}(t_0) = y^{(p-1)}(t_0)$$

there exist $t_1$ and $t_p$ such that

$$t_0 - \delta < t_1 < t_0 < t_p < t_0 + \delta$$

$$y(t_p) < x(t_p) < z_2(t_p)$$

and

$$z_2(t_1) < y(t_1) < x(t_1).$$

Again applying Corollary 15.2, we can assert that there is a $\varepsilon_2 > 0$ such that, if $z_3(t)$ is the solution of (9.1) with initial conditions

$$z_3^{(i)}(t_0) = z_2^{(i)}(t_0); \ 0 \le i \le n-1, \ i \ne p-3$$

$$z_3^{(p-3)}(t_0) = z_2^{(p-3)}(t_0) - \varepsilon_2 ,$$

then

$$y(t_0 \pm \delta) < z_3(t_0 \pm \delta) < x(t_0 \pm \delta)$$

$$y(t_p) < x(t_p) < z_3(t_p)$$

and

$$z_3(t_1) < y(t_1) < x(t_1).$$

Since

$$z_3^{(p-3)}(t_0) < z_2^{(p-3)}(t_0) = x^{(p-3)}(t_0) = y^{(p-3)}(t_0)$$

and p-3 is odd, it follows that there are $t_2$, $t_{p-1}$ such that

$$t_1 < t_2 < t_0 < t_{p-1} < t_p$$

$$z_3(t_{p-1}) < y(t_{p-1}) < x(t_{p-1})$$

and

$$y(t_2) < x(t_2) < z_3(t_2).$$

Proceeding in this way we obtain a solution $z(t) = z_{p/2+1}(t)$ and points

$$t_0 - \delta < t_1 < \ldots < t_{p/2} < t_0 < t_{p/2+1} < \ldots < t_p < t_0 + \delta$$

such that

$$y(t_0 \pm \delta) < z(t_0 \pm \delta) < x(t_0 \pm \delta)$$

$$z(t_0) = y(t_0) = x(t_0)$$

$$z(t_j) < y(t_j) < x(t_j); \ 1 \leq j \leq p, \ j \text{ odd}$$

and

$$y(t_j) < x(t_j) < z(t_j); \ 1 \leq j \leq p, \ j \text{ even}.$$

It follows that each of $z(t) - x(t)$ and $z(t) - y(t)$ has a zero on each of the intervals $[t_j, t_{j+1}]$, $1 \leq j \leq p-1$. Also, $z(t) - y(t)$ has a zero on $(t_0 - \delta, t_1)$ and $z(t) - x(t)$ has a zero on $(t_p, t_0 + \delta)$. Hence, each of $z(t) - x(t)$ and $z(t) - y(t)$ has p distinct zeros on $(t_0 - \delta, t_0 + \delta)$.

It is clear that if $[c,d] \subset (a,b)$ and $\varepsilon > 0$ were given, the construction of $z(t)$ could be carried out in such a way that either $|z(t) - x(t)| < \varepsilon$ on $[c,d]$ or $|z(t) - y(t)| < \varepsilon$ on $[c,d]$.

Corollary 17.2    Let the differential equation (9.1) be such that the conditions (A), (B), (C) and (D) of Theorem 15.5 are satisfied. Let $x(t), \cdot y(t)$ be solutions of (9.1) such that $x(t) - y(t)$ has a zero of order n-1 at some $t_0 \in (a,b)$. Then, $x(t) \neq y(t)$ for $t \neq t_0$.

Proof.    Assume that $x(t)$, $y(t)$ are solutions of (9.1) with a zero of order n-1 at $t_0$, and assume that $x(t_1) = y(t_1)$ for some $t_1 \neq t_0$. To be specific, assume that $t_1 > t_0$ and that $x(t) > y(t)$ on $(t_0, t_1)$. Then, by Theorem 17.1 there is a $\delta > 0$ and a solution $z(t)$ of (9.1) such that

$$a < t_0 - \delta < t_0 + \delta < t_1 < b$$

$$y(t_0 + \delta) < z(t_0 + \delta) < x(t_0 + \delta)$$

and such that each of $z(t) - x(t)$ and $z(t) - y(t)$ has n-1 distinct zeros on $(t_0 - \delta, t_0 + \delta)$. Since $z(t)$ extends throughout $(a,b)$, either $z(t) - x(t)$ or $z(t) - y(t)$ has n distinct zeros on $(a,b)$. This contradicts condition (D) and we conclude that $x(t) \neq y(t)$ for $t \neq t_0$.

**Corollary 17.3**   Let the differential equation (9.1) be such that the conditions (A), (B), (C) and (D) of Theorem 15.5 are satisfied. Then, each two point BVP in which data upto and including the (n-2)th derivative are assigned at one of the two points has at most one solution.

**Theorem 17.4**   Let the differential equation (9.1) be such that the conditions (A), (B), (C) and (D) of Theorem 15.5 are satisfied. Then, each r point BVP for (9.1) on $(a,b)$, $2 \leq r \leq n-1$ has at most one solution.

**Proof.**   Assume that the conclusion of Theorem 17.4 is false. Then, some r point BVPs, $2 \leq r \leq n-1$ have two or more distinct solutions. Let m be the largest integer for which this is the case. Then, $m < n$ and all (m+1) point BVPs have at most one solution.

Let $a < a_1 < \ldots < a_m < b$ be such that there exist distinct solutions $x(t)$ and $y(t)$ with $x(t) - y(t)$ having a zero of order $p_j \geq 1$ at $a_j$ for $1 \leq j \leq m$ where $\sum_{j=1}^{m} p_j \geq n$. We assume that for each $1 \leq j \leq m$ the order of the zero of $x(t) - y(t)$ at $a_j$ is exactly $p_j$. Since $m < n$ at least one of the zeros is a multiple zero.

Let $(\omega^-, \omega^+)$ be the maximal interval of existence of the solution $x(t)$ and choose $a_0$ and $a_{m+1}$ to satisfy $\omega^- < a_0 < a_1 < \ldots < a_m < a_{m+1} < \omega^+$. Concerning the zeros of $x(t) - y(t)$ we consider first the following two cases :

  (i)    $m = n-1$ and each $p_j$, $1 \leq j \leq m$ is an odd integer, or

  (ii)   $m < n-1$, there is a $k$, $1 \leq k \leq m$ such that $p_j = 1$ for $1 \leq j \leq m$, $j \neq k$ and such that $m + p_k$ and $n$ have the same parity.

In case (ii) let k be such that the stated conditions are satisfied and in case (i) let k be such that $p_k > 1$. Then in both cases $p_k > 1$ and $m - 1 + p_k \geq n$. Since $x(t)$ and $y(t)$ are distinct solutions of (9.1), we can always assume that $x^{(p_k)}(a_k) > y^{(p_k)}(a_k)$. Thus, from Theorem 17.1 and Corollary 15.2 we can assert that given any $\delta > 0$ such that $\delta < \frac{1}{2}$ min $\{a_j - a_{j-1} : 1 \leq j \leq m+1\}$ there is a solution $z(t)$ of (9.1) such that $[a_0, a_{m+1}]$ is contained in its maximal interval of existence, such that $z(t) - y(t)$ has $p_k$ distinct zeros in $(a_k - \delta, a_k + \delta)$, such that $z(t) - y(t)$ has at least one zero in $(a_j - \delta, a_j + \delta)$ for $1 \leq j \leq m$, $j \neq k$ but such that $z(t) \not\equiv y(t)$. Since $m - 1 + p_k \geq n$, this contradicts the assumption (D). Thus cases (i) and (ii) are ruled out.

In all other cases we can choose a k, $1 \leq k \leq m$ with $p_k > 1$ and integers $q_j$, $1 \leq j \leq m$ such that

$$1 \leq q_j \leq p_j ; \quad 1 \leq j \leq m, \ j \neq k$$

and

$$0 \leq q_k = p_k - 2s$$

where $s \geq 1$ is an integer and such that $\sum_{j=1}^{m} q_j = n-2$. We can assume that solutions $x(t)$, $y(t)$ are labelled so that $x^{(p_k)}(a_k) > y^{(p_k)}(a_k)$.

There is a $\varepsilon > 0$ such that $[a_0, a_{m+1}]$ is contained in the maximal interval of existence of any solution $z(t)$ of (9.1) which satisfies

$$|z^{(i)}(a_0) - x^{(i)}(a_0)| < \varepsilon, \quad 0 \leq i \leq n-1.$$

Let

$$\Delta \equiv \{(t_1, \ldots, t_{m+1}) : a_0 < t_1 < t_2 < \ldots < t_{m+1} < a_{m+1}\}$$

and

$$Q \equiv \{(c_0, \ldots, c_{n-1}) : |x^{(i)}(a_0) - c_i| < \varepsilon \text{ for } 0 \leq i \leq n-1\}.$$

Define the mapping $\phi : \Delta \times Q \to R^{n+m+1}$ by

$$\phi(t_1, \ldots, t_{m+1}, c_0, c_1, \ldots, c_{n-1})$$

$$= (t_1, \ldots, t_{m+1}, z(t_1), z'(t_1), \ldots, z^{(q_1-1)}(t_1), \ldots, z(t_k), \ldots,$$

$$z^{(q_k)}(t_k), \ldots, z(t_m), \ldots, z^{(q_m-1)}(t_m), z(t_{m+1}))$$

where in the image point $z(t)$ is the solution of (9.1) satisfying the initial conditions $z^{(i)}(a_0) = c_i$ for $0 \le i \le n-1$ and the last $n$ coordinates of the image point are, in the order of the $t_j$'s

$$z^{(i)}(t_j); \quad 1 \le j \le k-1, \quad 0 \le i \le q_j-1$$

$$z^{(i)}(t_k), \quad 0 \le i \le q_k$$

$$z^{(i)}(t_j); \quad k+1 \le j \le m, \quad 0 \le i \le q_j-1$$

and

$$z(t_{m+1}).$$

By the continuity of initial value problems with respect to initial conditions, $\phi$ is continuous on $\Delta \times Q$. Furthermore, since $(m+1)$ point BVPs for (9.1) have at most one solution on $(a,b)$ the mapping $\phi$ is one to one on $\Delta \times Q$. Since $\Delta \times Q$ is a connected open subset of $R^{n+m+1}$ it follows from Theorem 15.4 that $\phi(\Delta \times Q)$ is an open subset of $R^{n+m+1}$ and $\phi$ is a homeomorphism on $\Delta \times Q$.

Thus, if $a_m < \xi < a_{m+1}$ is fixed there is a $\varepsilon_0 > 0$ such that for $0 < \varepsilon < \varepsilon_0$ there is a solution $x_\varepsilon(t)$ of (9.1) such that $[a_0, a_{m+1}]$ is contained in the maximal interval of existence of $x_\varepsilon(t)$ and such that $x_\varepsilon(t)$ satisfies the conditions :

$$x_\varepsilon^{(i)}(a_j) = x^{(i)}(a_j); \quad 0 \le i \le q_j-1, \quad 1 \le j \le m, \quad j \ne k$$

$$x_\varepsilon^{(i)}(a_k) = x^{(i)}(a_k), \quad 0 \le i \le q_k-1 \quad \text{(omitted if } q_k = 0)$$

$$x_\varepsilon^{(q_k)}(a_k) = x^{(q_k)}(a_k) - \varepsilon$$

and

$$x_\varepsilon(\xi) = x(\xi).$$

This follows from the fact that $\phi(\Delta \times Q)$ is an open set and $(a_1, \ldots, a_m,$ $\xi, x(a_0), x'(a_0), \ldots, x^{(n-1)}(a_0)) \in \Delta \times Q$. Since $\phi^{-1}$ is continuous it follows that

$$\lim_{\varepsilon \to 0^+} x_\varepsilon(t) = x(t)$$

uniformly on $[a_0, a_{m+1}]$. Then, it follows from $q_k = p_k - 2s$ and $x^{(p_k)}(a_k) > y^{(p_k)}(a_k)$ that for $\varepsilon$ sufficiently small $y(t) - x_\varepsilon(t)$ has at least one zero on each of the open intervals $(a_{k-1}, a_k)$ and $(a_k, a_{k+1})$. In addition, $y(t) - x_\varepsilon(t)$ has a zero of order $q_j$ at $a_j$ for $1 \le j \le m$ where $\sum_{j=1}^{m} q_j = n-2$. Thus, $y(t) - x_\varepsilon(t)$ has at least $m+1$ distinct zeros in $(a,b)$ with the sum of the multiplicities of the zeros at least $n$. Since $r = m$ was assumed to be the largest value of $r$ for which some $r$ point BVPs have nonunique solutions, it follows that $y(t) \equiv x_\varepsilon(t)$ for sufficiently small $\varepsilon > 0$. But this is impossible since $y^{(q_k)}(a_k) = x^{(q_k)}(a_k) = x_\varepsilon^{(q_k)}(a_k) + \varepsilon$. From this contradiction we conclude that each $r$ point BVP for (9.1), $2 \le r \le n-1$ has at most one solution on $(a,b)$.

Theorem 17.5    Let the differential equation (9.1) be such that the conditions (A), (B) and (C) of Theorem 15.5 are satisfied. Further, let $h$ be an integer such that $2 \le h-1 < h < n$ and all $h$ point and all $(h-1)$ point BVPs for (9.1) on $(a,b)$ have at most one solution. Then, all $r$ point BVPs for (9.1) on $(a,b)$ with $2 \le r \le h$ have at most one solution.

Proof.    Assume that the conclusion of Theorem 17.5 is false.  Then,
h must satisfy $2 < h-1 < h < n$ and there is a largest integer m with
$2 \leq m < h-1$ such that a m point BVP for (9.1) on (a,b) has two distinct
solutions.  It follows from the maximality of m and the fact that
$m < h-1 < h$ that all (m+1) point and (m+2) point BVPs for (9.1) on (a,b)
have at most one solution.  Rest of the proof is same as that of Theorem
17.4 except for some minor changes.

**Theorem 17.6**   Let the differential equation (9.1) be such that the
conditions (A), (B) and (C) of Theorem 15.5 are satisfied.  Further, let
all (n-1) point BVPs for (9.1) on (a,b) have at most one solution.  Then,
all r point BVPs for (9.1) on (a,b) have at most one solution for each
r with $2 \leq r \leq n-1$.

Proof.    As a consequence of Theorem 17.5 it suffices to show that all
(n-2) point BVPs for (9.1) on (a,b) have at most one solution.  Assume
that this is not the case and that x(t) and y(t) are distinct solutions
of (9.1) such that x(t) - y(t) has a zero of exact order $p_j \geq 1$ at $a_j$
for each $1 \leq j \leq n-2$, where $a < a_1 < a_2 < ... < a_{n-2} < b$ and $\sum_{j=1}^{n-2} p_j \geq n$.

We consider two cases.  For the first case we assume that each $p_j$
is an odd integer or that one $p_j$ is an even integer and all other $p_j = 1$.
For either of these alternatives we let $1 \leq k \leq n-2$ be such that
$p_k = \{\max p_j : 1 \leq j \leq n-2\}$.  Then, $p_k \geq 3$ and as in Theorem 17.4, we
obtain a contradiction of the uniqueness of solutions of (n-1) point
BVPs.

In the second case, which is the complement of the first case,
there is a $k, 1 \leq k \leq n-2$ such that $p_k$ is even and such that $\sum_{\substack{j=1 \\ j \neq k}}^{n-2} p_j \geq n-2$.
It follows that we can choose $1 \leq q_j \leq p_j$; $1 \leq j \leq n-2$, $j \neq k$ such that

$$\sum_{\substack{j=1 \\ j \neq k}}^{n-2} q_j = n-2. \quad \text{Assume } x^{(p_k)}(a_k) > y^{(p_k)}(a_k) \text{ and let } \xi \text{ be chosen such}$$

that $a_{n-2} < \xi < b$ and such that $\xi$ is in the maximal interval of existence of the solution $x(t)$. Then, as in Theroem 17.4 for $\varepsilon > 0$ sufficiently small there is a solution $x_\varepsilon(t)$ of (9.1) such that

$$x_\varepsilon^{(i)}(a_j) = x^{(i)}(a_j); \; 0 \leq i \leq q_j - 1, \; 1 \leq j \leq n-2, \; j \neq k$$

$$x_\varepsilon(a_k) = x(a_k) - \varepsilon$$

$$x_\varepsilon(\xi) = x(\xi).$$

For $\varepsilon > 0$ sufficiently small, $x_\varepsilon(t) - y(t)$ has a zero in each of the
open intervals $(a_{k-1}, a_k)$ and $(a_k, a_{k+1})$, where $a_0 = a$ and $a_{n-1} = b$ and
$x_\varepsilon(t) - y(t)$ has a zero of order $q_j$ at $a_j$ for $1 \leq j \leq n-2$, $j \neq k$. This
again contradicts the uniqueness of solutions of (n-1) point BVPs and
the theorem is proved.

Next, we state a result for the right focal point BVPs which has
been proved in [2] following the similar argument as that of Theorem
17.4.

Theorem 17.7    Let the differential equation (9.1) be such that the
conditions (A), (B) and (C) of Theorem 15.5 are satisfied. Further, let
all right $(1,\ldots,1)$ focal point BVPs for (9.1) on $(a,b)$ have at most
one solution. Then, given $r$, $2 \leq r \leq n-1$ all right $(m_1,\ldots,m_r)$ focal
point BVPs for (9.1) on $(a,b)$ have at most one solution.

Remark 17.1    If the statements concerning BVPs are assumed to be the
points at which data is specified are interior points of the maximal
interval of existence of the solutions, then in all the above results
hypothesis (B) can be dropped.

So far our results prove that the uniqueness of the BVPs with the boundary conditions prescribed at m points implies the uniqueness of some BVPs with boundary conditions prescribed at r ( < m) points. For the linear differential equation (2.1) there has been considerable effort to study the converses of the results of this type. In 1960 in one of the first such results Azbelev and Tsalyuk [1] proved for n = 3 that if all two point BVPs have unique solutions on (a,b), then all three point BVPs have unique solutions on (a,b). In 1965 Sherman [9] proved the validity of the corresponding result for arbitrary n. For the nonlinear equation (9.1) the known result for n = 3 is stated in the following :

**Theorem 17.8**   Let n = 3 and the differential equation (9.1) be such that the conditions (A) and (B) of Theorem 15.5 are satisfied. Further, let all two point BVPs for (9.1) on (a,b) have at most one solution. Then, all three point BVPs for (9.1) on (a,b) have at most one solution.

**Proof.**   Assume that the conclusion of Theorem 17.8 is false. Then, there are points $a < a_1 < a_2 < a_3 < b$ and solutions $x(t)$, $y(t)$ of (9.1) such that $x(a_i) = y(a_i)$ for $i = 1,2$ and 3 with $x(t) \not\equiv y(t)$ on $[a_1, a_3]$. In view of our hypotheses it follows that $x'(a_i) \neq y'(a_i)$ for each $i = 1,2,3$. Hence, without loss of generality we can assume $x(t) > y(t)$ on $(a_1, a_2)$ and $x(t) < y(t)$ on $(a_2, a_3)$.

For each $m \geq 1$, let $x_m(t)$ be a solution on (a,b) of (9.1) satisfying the initial conditions

$$x_m(a_1) = x(a_1), \quad x_m'(a_1) = x'(a_1) \text{ and } x_m''(a_1) = x''(a_1) + m.$$

Then, from the uniqueness of two point BVPs it follows that for each $m \geq 1$

$$x_{m+1}(t) > x_m(t) > x(t)$$

for all $t \neq a_1$ in (a,b). For each $m \geq 1$ let

$$E_m = \{t : a_2 \leq t \leq a_3 \text{ and } x_m(t) \leq y(t)\}.$$

From our hypotheses it is not difficult to see that the sets $E_m$ must be nonempty for each $m \geq 1$. Thus, $E_{m+1} \subset E_m \subset (a_2, a_3)$ for each $m \geq 1$ and each $E_m$ is nonempty and compact. It follows that $\bigcap\limits_{m=1}^{\infty} E_m = E \neq \phi$.

Next, we observe that the set $E$ consists of a single point $t_0$ with $a_2 < t_0 < a_3$. In fact, if $t_1, t_2 \in E$ with $a_2 < t_1 < t_2 < a_3$, then the same type of argument that one uses to show that the foregoing sets $E_m$ are nonempty leads to the conclusion that the interval $[t_1, t_2]$ must be contained in $E$. However, $[t_1, t_2] \subset E$ implies that the sequence $\{x_m(t)\}$ is uniformly bounded on $[t_1, t_2]$ which contradicts Theorem 16.5. Thus, we conclude that $E = \{t_0\}$ with $a_2 < t_0 < a_3$ and $\lim\limits_{m \to \infty} x_m(t_0) = x_0 \leq y(t_0)$.

Now we claim this is not possible. First, assume $x_0 = y(t_0)$. Then, for each $\varepsilon > 0$ sufficiently small there is a solution $y(t; \varepsilon)$ of (9.1) such that $y(a_1; \varepsilon) = y(a_1)$, $y'(a_1; \varepsilon) = y'(a_1)$, $y(t_0; \varepsilon) = y(t_0) - \varepsilon$ and $y(t; \varepsilon) \leq y(t)$ on $[a_1, a_3]$. Such a solution $y(t; \varepsilon)$, where $\varepsilon$ is chosen so that $y(t_0; \varepsilon) = y(t_0) - \varepsilon > x(t_0)$ can be used in place of $y(t)$ in defining the sets $\{E_m\}$ with respect to the given sequence of solutions $\{x_m(t)\}$. Then as previously, it would follow that each of these sets would be nonempty which is impossible.

The remaining possibility is that $x(t_0) < x_0 < y(t_0)$. In this case, since the line segment consisting of the points with coordinates

$$\lambda x^{(i)}(a_1) + (1-\lambda)y^{(i)}(a_1)$$

where $i = 0, 1, 2$ and $0 \leq \lambda \leq 1$ is a connected set of $R^3$, there is a $\lambda_0$, $0 \leq \lambda_0 \leq 1$ and a solution $x(t; \lambda_0)$ of (9.1) such that

$$x^{(i)}(a_1; \lambda_0) = \lambda_0 x^{(i)}(a_1) + (1-\lambda_0)y^{(i)}(a_1)$$

for $i = 0,1,2$ and such that $x(t_0;\lambda_0) = x_0$. Of course, we have $x(a_1;\lambda_0)$ $= x(a_1) = y(a_1)$. Now, there is a $\eta > 0$ such that $[t_0-\eta,t_0+\eta] \subset (a_2,a_3)$ and such that $x(t;\lambda_0) < y(t)$ on $[t_0-\eta,t_0+\eta]$. Then, with $\{x_m(t)\}$ the same sequence as previously we have

$$\lim x_m(t) > y(t) > x(t;\lambda_0)$$

for all $t \neq t_0$ in $[t_0-\eta,t_0+\eta]$ and

$$\lim x_m(t_0) = x_0 = x(t_0;\lambda_0).$$

This is the same contradictory situation as the case $x_0 = y(t_0)$ considered previously.

From this final contradiction we conclude that $x_0 \leq y(t_0)$ is impossible and hence the conclusion of Theorem 17.8 follows.

## COMMENTS AND BIBLIOGRAPHY

Theorem 17.1 is taken from [6], also for an alternative proof see [7]. Theorem 17.4 is due to Jackson [5]. Theorems 17.5 and 17.6 are given in [3] and cover some results for the linear equations proved in [8]. Theorem 17.8 is adapted from Jackson [4].

1.    Azbelev, N. and Tsalyuk, Z. "On the question of the distribution of the zeros of solutions of a third order linear differential equation", Mat. Sb. (N.S.) 51, 475-486 (1960).
2.    Henderson, J. "Uniqueness of solutions of right focal point boundary value problems for ordinary differential equations", J. Diff. Equs. 41, 218-227 (1981).
3.    Henderson, J. and Jackson, L. "Existence and uniqueness of solutions of k-point boundary value problems for ordinary differential equations", J. Diff. Equs. 48, 373-385 (1983).
4.    Jackson, L. K. "Existence and uniqueness of solutions of boundary value problems for third order differential equations", J. Diff.

Equs. <u>13</u>, 432-437 (1973).

5.  Jackson, L. K. "Uniqueness of solutions of boundary value problems
    for ordinary differential equations", SIAM J. Appl. Math. <u>24</u>,
    535-538 (1973).

6.  Jackson, L. and Klaasen, G. "Uniqueness of solutions of boundary
    value problems for ordinary differential equations", SIAM J. Appl.
    Math. <u>19</u>, 542-546 (1970).

7.  Klaasen, G. "Existence theorems for boundary value problems for
    nth order ordinary differential equations", Rocky Mountain J.
    Math. <u>3</u>, 457-472 (1973).

8.  Peterson, A. C. "A theorem of Aliev", Proc. Amer. Math. Soc. <u>23</u>,
    364-366 (1969).

9.  Sherman, T. "Properties of solutions of nth order linear differen-
    tial equations", Pacific J. Math. <u>15</u>, 1045-1060 (1965).

## 18. BOUNDARY VALUE FUNCTIONS

The concept of boundary value functions for the third order linear differential equations appears to have been used first by Azbelev and Tsalyuk [1], whereas for the nth order linear differential equations by Sherman [11]. Peterson [3] was the first to extend this notion to non-linear differential equation (9.1).

Throughout, we shall always assume that the differential equation (9.1) satisfies the conditions (A), (B) and (C) of Corollary 15.2 on [a,b], where $b \leq \infty$. Before we define these boundary value functions for (9.1), we state the preliminary definition.

**Definition 18.1** We say that $x \in C^{(n)}[a,b]$ has a $(\lambda(1),\ldots,\lambda(r))$ distribution of zeros, $0 \leq \lambda(i) \leq n$, $\sum_{i=1}^{r} \lambda(i) = n$, on $[c,d] \subset [a,b]$ provided there are points $c \leq t_1 < \ldots < t_r \leq d$ such that $x(t)$ has a zero of order at least $\lambda(i)$ at $t_i$, $1 \leq i \leq r$.

**Definition 18.2** Let $s \in [a,b]$ and $C(s) = \{c > s$ : there are distinct solutions $x(t)$ and $y(t)$ of (9.1) such that $x(t) - y(t)$ has a $(\lambda(1),\ldots,\lambda(r))$ distribution of zeros on $[s,c]\}$. The function

$$c_{\lambda(1)\ldots\lambda(r)}(s) = \begin{cases} \inf C(s) & \text{if } C(s) \neq \phi \\ \infty & \text{if } C(s) = \phi \end{cases}$$

is called a boundary value function.

**Remark 18.1** From the hypothesis (C) it follows that $c_n(s) = \infty$. Further, if $s \leq a_1 < \ldots < a_r < c_{\lambda(1)\ldots\lambda(r)}(s) \leq \infty$, then the solutions of the BVP (9.1), (2.4) where $k_i = \lambda(i)-1$, $1 \leq i \leq r$ when exist are unique.

**Definition 18.3**   The first conjugate point $\eta_1(s)$ for the differential equation (9.1) is defined by $\eta_1(s) = \min\{c_{\lambda(1)\ldots\lambda(r)}(s) : \sum_{i=1}^{r} \lambda(i) = n\}$.

**Definition 18.4**   If J is an interval and $J \subset [a,b)$, then we say that J is an interval of disconjugacy for (9.1) provided there do not exist distinct solutions $x(t)$, $y(t)$ of (9.1) such that $x(t) - y(t)$ has at least n zeros, counting multiplicities, on J.   From the Definition 18.3 it follows that if $J \subset [s,\eta_1(s))$, then J is an interval of disconjugacy and that $[s,\eta_1(s))$ is a maximal half open interval of disconjugacy.

**Remark 18.2**   From Theorem 17.4 it follows that $\eta_1(s) = c_{\lambda(1)\ldots\lambda(r)}(s)$. Thus, if the function $f(t,\underline{x})$ is linear or satisfies the Lipschitz condition (9.12) on each $[c,d] \times R^{q+1}$, where $[c,d]$ is a compact subinterval of $[a,b)$, then Theorem 9.14 ensures that $\eta_1(s) > s$.   However, in general this is not true.   For example, consider the differential equation $x'' = -x^3$ and notice that $x \equiv 0$ is a solution.   Obviously, conditions (A) - (C) on $[0,\infty)$ are satisfied.   It is easy to show that for $\varepsilon > 0$ given there is a nontrivial solution with two zeros in $[0,\varepsilon)$, and so $0 \le \eta_1(0) \le \varepsilon$ for each $\varepsilon > 0$ and hence $\eta_1(0) = 0$.   Since the equation is autonomous $\eta_1(s) = s$.

**Definition 18.5**   Let $s \ge a$.   If $c_{\lambda(1)\ldots\lambda(r)}(s_1) = \infty$ for all $s_1 > s$, then we write $c_{\lambda(1)\ldots\lambda(r)}(s+) = \infty$.   Otherwise $c_{\lambda(1)\ldots\lambda(r)}(s+)$ is defined by $c_{\lambda(1)\ldots\lambda(r)}(s+) = \inf\{c_{\lambda(1)\ldots\lambda(r)}(s_1) : s_1 > s\}$.

**Definition 18.6**   Let $x(t)$ be a solution of (9.1), and let $s \le t_1 < \ldots < t_r < c_{\lambda(1)\ldots\lambda(r)}(s)$.   Define

$$S(x; t_1^{\lambda(1)},\ldots, \bar{t}_k^{\lambda(k)},\ldots, t_r^{\lambda(r)}) \equiv$$

$$\{y^{(\lambda(k)-1)}(t_k) : y(t) \text{ is a solution of (9.1) such that}$$

$$y^{(j)}(t_i) = x^{(j)}(t_i); \ 1 \le i \le r, \ i \ne k, \ 0 \le j \le \lambda(i) - 1$$

$$y^{(j)}(t_k) = x^{(j)}(t_k), \ 0 \le j \le \lambda(k) - 2 \quad \text{if } \lambda(k) \ge 2 \}.$$

**Remark 18.3**  If $n = 2$ and $s \le a_1 < a_2 < \eta_1(s)$, then from Theorem 15.3 for any solution $x(t)$ of (7.1) $S(x;a_1,\bar{a}_2) = R$ which is equivalent to the existence of solutions of all two point BVPs (7.1), (7.2). Similarly, if condition (E) of Theorem 15.7 on $[a,b]$ is satisfied and $s \le a_1 < \ldots < a_r < \eta_1(s)$, then from Theorem 15.8 for any solution $x(t)$ of (9.1) $S(x;a_1^{\lambda(1)},\ldots,\bar{a}_k^{\lambda(k)},\ldots,a_r^{\lambda(r)}) = R, \ 1 \le k \le r.$

**Definition 18.7**  Let $x_0(t)$ be a solution of (9.1), then the linear differential equation

$$(18.1) \qquad y^{(n)} = \sum_{i=0}^{q} f_i(t,x_0(t),\ldots,x_0^{(q)}(t))y^{(i)}$$

is called the variational equation of (9.1) along the solution $x_0(t)$. In (18.1) $f_i(t,x,\ldots,x^{(q)}) = \dfrac{\partial f}{\partial x^{(i)}} (t,x,\ldots,x^{(q)})$ are assumed to be continuous on $[a,b] \times R^{q+1}$.

**Definition 18.8**  We will denote the first conjugate point of $t = s$ for (18.1) by $\eta_1(s,x_0(t))$.

In Theorems 18.2 - 18.5, we shall provide several ordering relations involving boundary value functions $c_{\lambda(1)\ldots\lambda(r)}(s)$. These relations are basically "uniqueness implies uniqueness" type of results and as has been noted in Section 17 are useful in proving "uniqueness implies existence" type of results. To prove these theorems, we need

**Lemma 18.1**  Assume $s < c_{\lambda(1)\ldots\lambda(r)}(s) \le \infty$, fix $t_1 \epsilon [s, c_{\lambda(1)\ldots\lambda(r)}(s))$, and let $\Delta = \{(t_2,\ldots,t_r) : s \le t_1 < \ldots < t_r < c_{\lambda(1)\ldots\lambda(r)}(s)\}.$

Define $\phi : \Delta \times R^{n-\lambda(1)} \to R^{r+n-\lambda(1)-1}$ by

$$\phi(t_2,\ldots,t_r,x_{\lambda(1)+1},\ldots,x_n)$$

$$= (t_2,\ldots,t_r,x(t_2),\ldots,x^{(\lambda(2)-1)}(t_2),\ldots,x(t_r),\ldots,x^{(\lambda(r)-1)}(t_r))$$

where $x(t)$ is the solution of (9.1) satisfying

$$x^{(j-1)}(t_1) = \begin{cases} \alpha_j & j = 1,2,\ldots,\lambda(1) \\ \\ x_j & j = \lambda(1)+1,\ldots,n \end{cases}$$

and $\alpha_1,\ldots,\alpha_{\lambda(1)}$ are fixed constants. Then, the mapping $\phi$ from $\Delta \times R^{n-\lambda(1)}$ onto $\phi(\Delta \times R^{n-\lambda(1)})$ is a homeomorphism and $\phi(\Delta \times R^{n-\lambda(1)})$ is open.

Proof.   The proof is similar to that of Theorem 15.5.

Theorem 18.2   If $q \geq 3$ and $p \geq 2$, then

$$c_{pq}(s) \geq \min \{ c_{p-1,q+1}(s),\ c_{p,1,q-1}(s),\ c_{p,1,q-2,1}(s) \}.$$

Also, if $q \geq 2$ and $p \geq 3$, then

$$c_{pq}(s+) \geq \min \{ c_{p+1,q-1}(s+),\ c_{p-1,1,q}(s+),\ c_{1,p-2,1,q}(s+) \}.$$

Proof.   We will prove only the first result since the proof of the second is similar.   Let $c = \min\{c_{p-1,q+1}(s),c_{p,1,q-1}(s),c_{p,1,q-2,1}(s)\}$ and suppose $c_{pq}(s) < c$.   Then, there exist two distinct solutions $x(t)$ and $y(t)$ of (9.1) such that $x(t) - y(t)$ has a $(p,q)$ distribution of zeros on $[s,c)$ at the points $t_1,t_2$ where $s \leq t_1 < t_2 < c$.   Since $c_{p-1,q+1}(s) \geq c$, the zero at $t_2$ is of order exactly $q$, and since $c_{p,1,q-1}(s) \geq c$ we can assume that $x(t) > y(t)$ for $t \in (t_1,t_2)$.

Let $\Delta = \{(s_2, s_3) : s \leq t_1 < s_2 < s_3 < c\}$ and define $\phi : \Delta \times R^{n-p} \to R^{n+2-p}$ by

$$\phi(s_2, s_3, x_{p+1}, \ldots, x_n) = (s_2, s_3, z(s_2), z(s_3), z'(s_3), \ldots, z^{(q-2)}(s_3))$$

where $z(t)$ is the solution of (9.1) satisfying $z^{(j)}(t_1) = x^{(j)}(t_1)$, $0 \leq j \leq p-1$ and $z^{(j-1)}(t_1) = x_j$, $p+1 \leq j \leq n$. By Lemma 18.1, since $s_3 < c \leq c_{p,1,q-1}(s)$, it follows that $\phi(\Delta \times R^{n-p})$ is an open subset of $R^{n+2-p}$ and $\phi$ mapping $\Delta \times R^{n-p}$ onto $\phi(\Delta \times R^{n-p})$ is a homeomorphism.

Now if $t_1 < \xi < t_2$, $(\xi, t_2, x(\xi), x(t_2), \ldots, x^{(q-2)}(t_2))$ is in $\phi(\Delta \times R^{n-p})$ and since $\phi(\Delta \times R^{n-p})$ is open, given $\varepsilon > 0$ and sufficiently small there exists a solution $x_1(t)$ of (9.1) satisfying

$$x_1^{(j)}(t_1) = x^{(j)}(t_1), \ 0 \leq j \leq p-1$$

$$x_1(\xi) = x(\xi)$$

$$x_1^{(j)}(t_2) = x^{(j)}(t_2), \ 0 \leq j \leq q-3$$

$$x_1^{(q-2)}(t_2) = x^{(q-2)}(t_2) + (-1)^{q+1}\varepsilon.$$

Further, it follows from the fact that $\phi^{-1}$ is continuous and the continuous dependence of solutions of initial conditions that $\lim_{\varepsilon \to 0^+} x_1(t) = x(t)$ uniformly on compact subsets of $[s, \infty)$, and so for $\varepsilon > 0$ sufficiently small $x_1(t) - y(t)$ has a zero of order at least $p$ at $t_1$, an odd order zero on $(t_1, t_2)$, a zero at $t_2$ of order exactly $q-2$ and a zero in $(t_2, c)$. That is, $x_1(t) - y(t)$ has a $(p, 1, q-2, 1)$ distribution of zeros on $[s, c)$, yet since $x_1^{(q-2)}(t_2) = y^{(q-2)}(t_2) + (-1)^{q+1}\varepsilon$, $x_1(t) \not\equiv y(t)$. This is the desired contradiction and so the theorem is proved.

Theorem 18.3   If $q \geq 2$, then

$$c_{p,1,q}(s) \geq \min \{ c_{p,q+1}(s), c_{p,1,1,q-1}(s) \} .$$

Also, if $p \geq 2$

$$c_{p,1,q}(s+) \geq \min \{ c_{p+1,q}(s+), c_{p-1,1,1,q}(s+) \} .$$

Proof. We prove only the first statement as the proof of the other is similar. Assume the statement is false and let $c = \min \{ c_{p,q+1}(s), c_{p,1,1,q-1}(s) \}$. Then, there are two distinct solutions $x(t)$ and $y(t)$ of (9.1) such that $x(t) - y(t)$ has a $(p,1,q)$ distribution of zeros at $t_1, t_2, t_3$ where $s \leq t_1 < t_2 < t_3 < c$. Also, $x(t) - y(t)$ has exactly one distinct zero in $(t_1, t_3)$ since $c_{p,1,1,q-1}(s) \geq c$, and the zero at $t_3$ is of order exactly $q$ since $c_{p,q+1}(s) \geq c$, so assume $x^{(q)}(t_3) > y^{(q)}(t_3)$.

Suppose first that the zero at $t_2$ is of odd exact order. Since $c_{p,q+1}(s) \geq c$ it follows from Lemma 18.1 that for $\varepsilon > 0$ sufficiently small that there exists a solution $x_\varepsilon(t)$ of (9.1) satisfying

$$x_\varepsilon^{(j)}(t_1) = x^{(j)}(t_1), \ 0 \leq j \leq p-1$$

$$x_\varepsilon^{(j)}(t_3) = x^{(j)}(t_3); \ 0 \leq j \leq q, \ j \neq q-1$$

$$x_\varepsilon^{(q-1)}(t_3) = x^{(q-1)}(t_3) + \varepsilon .$$

Furthermore, by Lemma 18.1 it follows that $\lim_{\varepsilon \to 0^+} x_\varepsilon(t) = x(t)$ uniformly on compact subintervals of $(s,c)$. Therefore, for $\varepsilon > 0$ sufficiently small $x_\varepsilon(t) - y(t)$ has a zero of order $p$ at $t_1$, two distinct zeros in $(t_1, t_3)$ and a zero of order $q-1$ at $t_3$. Thus contradicting the assumption that $c_{p,1,1,q-1}(s) \geq c$. Hence in this case the theorem is valid.

Now suppose the zero at $t_2$ is of even exact order. Proceeding as above for $\eta > 0$ and sufficiently small there exists by Lemma 18.1 a solution $x_\eta(t)$ of (9.1) satisfying

$$x_\eta^{(j)}(t_1) = x^{(j)}(t_1), \ 0 \le j \le p-1$$

$$x_\eta^{(j)}(t_3) = x^{(j)}(t_3), \ 0 \le j \le q-1$$

$$x_\eta^{(q)}(t_3) = x^{(q)}(t_3) - \eta.$$

Then, for $\eta > 0$ sufficiently small we have that $x_\eta(t) - y(t)$ and $x_\eta(t) - x(t)$ both have zeros at $t_1$ of orders at least p and at $t_3$ of order q, and yet $x_\eta(t) \not\equiv y(t)$ and $x_\eta(t) \not\equiv x(t)$. If either $x_\eta(t) - x(t)$ or $x_\eta(t) - y(t)$ has an odd order zero in $(t_1, t_3)$ we are in fact back to the previous case.

Hence, $x_\eta(t) - y(t)$ for $\eta > 0$ sufficiently small has in addition an even order zero at $t_2$, and $x_\eta(t)$ lies between $x(t)$ and $y(t)$ in $(t_1, t_2) \cup (t_2, t_3)$. Fix such a $\eta > 0$. Now, we again apply Lemma 18.1 to $c_{p,q+1}(s)$ to obtain for $\delta > 0$ sufficiently small the existence of a solution $x_\delta(t)$ of (9.1) satisfying

$$x_\delta^{(j)}(t_1) = x_\eta^{(j)}(t_1), \ 0 \le j \le p-1$$

$$x_\delta^{(j)}(t_3) = x_\eta^{(j)}(t_3); \ 0 \le j \le q, \ j \ne q-1$$

$$x_\delta^{(q-1)}(t_3) = x_\eta^{(q-1)}(t_3) + \delta.$$

Note that $\lim_{\delta \to 0^+} x_\delta(t) = x_\eta(t)$ uniformly on $[t_1, t_3]$. Now, $x_\delta(t) - y(t)$ and $x_\delta(t) - x(t)$ both have zeros of order p and q-1 at $t_1$ and at $t_3$ respectively, and yet $x_\delta(t) \not\equiv x(t)$ and $x_\delta(t) \not\equiv y(t)$. Also, for $\delta > 0$ sufficiently small $x_\delta(t) - y(t)$ has a zero near $t_3$. Since $c_{p,1,1,q-1}(s) \ge c$, $x_\delta(t) - y(t)$ has no other zeros in $(t_1, t_3)$. It then follows that $x_\delta(t) - x(t)$ has a (p,1,1,q-1) distribution of zeros on $[t_1, t_3]$ which is a contradiction.

Theorem 18.4    The following hold

221

(i)   $c_{n-1,1}(s) \geq c_{n-2,1,1}(s) \geq \cdots \geq c_{2,1,\ldots,1}(s) \geq c_{1\ldots1}(s)$

(ii)  $c_{\lambda(1)\ldots\lambda(r)}(s) \geq \min\{c_{k,1,\ldots,1}(s), c_{k,n-k}(s)\}$, for $\lambda(1) \geq k$.

**Proof.** The proof uses techniques of Theorems 17.4, 18.2 and 18.3.

**Theorem 18.5**   Assume that the condition (E) of Theorem 15.7 on $[a,b]$ is satisfied. Then, the following hold

(i)   $c_{n-2,1,1}(s) \geq \min\{c_{n-1,1}(s), c_{n-2,2}(s)\}$

(ii)  $c_{1,1,n-2}(s) \geq \min\{c_{1,n-1}(s), c_{2,n-2}(s)\}$.

**Proof.** The proof uses techniques of Theorems 17.8, 18.2 and 18.3.

Next, we shall prove some more "uniqueness implies existence" type of results. For this, we begin with

**Lemma 18.6**   Assume $a \leq s < c_{\lambda(1)\ldots\lambda(r)} \leq \infty$, and let $\Delta = \{(t_1,\ldots,t_r) \in R^r: s \leq t_1 < \cdots < t_r < c_{\lambda(1)\ldots\lambda(r)}(s)\}$.

Define $\phi : \Delta \times R^n \to R^{r+n}$ by

$$\phi(t_1,\ldots,t_r,x_1,\ldots,x_n)$$

$$= (t_1,\ldots,t_r,x(t_1),x'(t_1),\ldots,x^{(\lambda(1)-1)}(t_1),x(t_2),\ldots,$$

$$x(t_r),\ldots,x^{(\lambda(r)-1)}(t_r))$$

where $x(t)$ is the solution of (9.1) satisfying $x^{(j-1)}(s) = x_j$, $1 \leq j \leq n$. Then, $\phi : \Delta \times R^n \to \phi(\Delta \times R^n)$ is a homeomorphism and $\phi(\Delta \times R^n)$ is a relatively open subset of $\Delta \times R^n$.

**Proof.** Let $\mu = s$ and $\nu = c_{\lambda(1)\ldots\lambda(r)}(s)$. Set $\bar{\Delta} : \{(t_1,\ldots,t_r) \in R^r : 2\mu - t_2 < t_1 < \nu$ and $\mu < t_2 < \cdots < t_r < \nu\}$. Define $\bar{\phi} : \bar{\Delta} \times R^n \to R^{r+n}$ by

$$\bar{\phi}(t_1,\ldots,t_r,x_1,\ldots,x_n)$$

$$= (t_1,\ldots,t_r,x(t_1),x'(t_1),\ldots,x^{(\lambda(1)-1)}(t_1),x(t_2),\ldots,$$

$$x(t_r),\ldots,x^{(\lambda(r)-1)}(t_r))$$

if $(t_1,\ldots,t_r,x_1,\ldots,x_n) \in \Delta \times R^n$, and

$$\bar{\phi}(t_1,\ldots,t_r,x_1,\ldots,x_n)$$

$$= (t_1,\ldots,t_r,x(2\mu-t_1),x'(2\mu-t_1),\ldots,x^{(\lambda(1)-1)}(2\mu-t_1),$$

$$x(t_2),x'(t_2),\ldots,x(t_r),\ldots,x^{(\lambda(r)-1)}(t_r))$$

if $(t_1,\ldots,t_r,x_1,\ldots,x_n) \in \bar{\Delta} \times R^n \sim \Delta \times R^n$, where $x(t)$ is the solution of (9.1) satisfying $x^{(j-1)}(s) = x_j$, $1 \leq j \leq n$.

Note that if $(t_1,\ldots,t_r,x_1,\ldots,x_n) \in \bar{\Delta} \times R^n \sim \Delta \times R^n$, then $2\mu - t_2 < t_1 < \mu < t_2 < \ldots < t_r < \nu$ , and hence $\mu < 2\mu - t_1 < t_2 < \ldots < t_r < \nu$ . The function $\bar{\phi}$ is one to one since, if two points of $\bar{\Delta} \times R^n$ have the same image under $\bar{\phi}$, then the first r coordinates of the two points must agree, and hence by the uniqueness of solutions of the BVP (9.1), (2.4) where $k_i = \lambda(i)-1$ the remaining n coordinates of these two points must agree. It follows from the continuous dependence of solutions of initial conditions that $\bar{\phi}$ is continuous. It is clear that $\bar{\Delta} \times R^n$ is an open subset of $R^{r+n}$, hence by Theorem 15.4, $\bar{\phi} : \bar{\Delta} \times R^n \to R^{r+n}$ is an open mapping. By the way, we have defined $\bar{\phi}$, it follows that $\phi(\Delta \times R^n) = \bar{\phi}(\bar{\Delta} \times R^n) \cap [\mu,\nu] \times R^{r+n-1}$, hence $\phi(\Delta \times R^n)$ is a relatively open subset of $[\mu,\nu] \times R^{r+n-1}$ and $\phi : \Delta \times R^n$ onto $\phi(\Delta \times R^n)$ is a homeomorphism.

**Corollary 18.7** Let $a \leq s \leq a_1 < \ldots < a_r < c_{\lambda(1)\ldots\lambda(r)}$, and let $x(t)$ be a solution of the BVP (9.1), (2.4) where $k_i = \lambda(i)-1$. Then, for $\delta > 0$ sufficiently small, $s \leq t_1 < \ldots < t_r < c_{\lambda(1)\ldots\lambda(r)}$, $|\delta_{j+1,i}| < \delta$, $|t_i - a_i| < \delta$; $1 \leq i \leq r$, $0 \leq j \leq \lambda(i)-1$, there exists a unique solution $y(t)$ of (9.1) satisfying

$$y^{(j)}(t_i) = A_{j+1,i} + \delta_{j+1,i} \; ; \; 1 \le i \le r, \; 0 \le j \le \lambda(i)-1.$$

Furthermore, given any compact subset $J \subset [a,b)$ and $\varepsilon > 0$, $\delta$ may be picked small enough so that $|y^{(i)}(t) - x^{(i)}(t)| < \varepsilon$, $0 \le i \le n-1$ for all $t \in J$.

Proof. The function $\phi$ defined in Lemma 18.6 is an open map, and $\phi(\Delta \times R^n)$ is a relatively open subset of $\Delta \times R^n$. Since the point $(a_1,\ldots,a_r, A_{1,1},$ $\ldots, A_{\lambda(1),1}, A_{1,2}, \ldots, A_{\lambda(r),r})$ is in the open set $\phi(\Delta \times R^n)$, there exists an open neighborhood, relative to $\Delta \times R^n$, contained in $\phi(\Delta \times R^n)$. Hence for $\delta$ sufficiently small, the set

$$S \equiv \{(t_1,\ldots,t_r, A_{1,1}+\delta_{1,1}, \ldots, A_{\lambda(r),r} + \delta_{\lambda(r),r}) :$$

$$s \le t_1 < \ldots < t_r < c_{\lambda(1)\ldots\lambda(r)}(s), \; |t_i - a_i| < \delta,$$

$$|\delta_{j+1,i}| < \delta; \; 1 \le i \le r, \; 0 \le j \le \lambda(i)-1\}$$

is contained in $\phi(\Delta \times R^n)$. By the way, $\phi$ is defined for each point of $S$ there exists a corresponding solution of equation (9.1) which determines that point. The last statement of the conclusion of this corollary is a consequence of the continuity of $\phi^{-1}$ and the continuous dependence of solutions on initial conditions.

Remark 18.4  Corollary 18.7 shows that uniqueness of the BVP (9.1), (2.4) where $k_i = \lambda(i)-1$ on an interval $[s, c_{\lambda(1)\ldots\lambda(r)}(s))$ implies "local" existence of solutions of the BVP (9.1), (2.4) where $k_i = \lambda(i)-1$.

Theorem 18.8  Let $1 \le k \le r$ and $s \le t_1 < \ldots < t_r < c_{\lambda(1)\ldots\lambda(r)}(s)$, where $\lambda(k) = 1$. If (9.1) satisfies condition (E) of Theorem 15.7 on $[a,b)$, then for any solution $x(t)$ of (9.1), $S(x; t_1^{\lambda(1)}, \ldots, \bar{t}_k, \ldots, t_r^{\lambda(r)})$ is an open interval.

Proof.  It follows from Corollary 18.7 that the set $S = S(x; t_1^{\lambda(1)}, \ldots, \bar{t}_k,$

$\dots, t_r^{\lambda(r)})$ is an open subset of R. It suffices to establish that if $\sigma, \tau \in S$ with $\sigma < \tau$, then $[\sigma, \tau] \subset S$. Assume that $\sigma, \tau \in S$ and let $\gamma_0 = \sup\{\gamma \leq \tau : [\sigma, \gamma] \subset S\}$. By Corollary 18.7, if $\gamma_0 \in S$, then $[\sigma, \gamma_0 + \varepsilon) \subset S$ for $\varepsilon > 0$ sufficiently small, contrary to the definition of $\gamma_0$. Hence $\gamma_0 \notin S$. If $\gamma_0 < \tau$, let $\{\gamma_m\}$ be a sequence of real numbers such that for each m, $\sigma < \gamma_m < \gamma_{m+1} < \gamma_0$ and $\{\gamma_m\}$ converges to $\gamma_0$. Let $\{y_m(t)\}$ be the corresponding sequence of solutions of (9.1) satisfying

$$y_m^{(j)}(t_i) = x^{(j)}(t_i); \ 1 \leq i \leq r, \ i \neq k, \ 0 \leq j \leq \lambda(i)-1$$

$$y_m(t_k) = \gamma_m.$$

Let $z(t)$ be the solution of (9.1) satisfying

$$z^{(j)}(t_i) = x^{(j)}(t_i); \ 1 \leq i \leq r, \ i \neq k, \ 0 \leq j \leq \lambda(i)-1$$

$$z(t_k) = \tau.$$

Note that $y_m(t) < y_{m+1}(t)$ for $t \in (t_k, t_{k+1})$ for each m (where we let $t_{r+1} = c_{\lambda(1)\dots\lambda(r)}(s)$ in case $k = r$). Similarly, $y_m(t) < z(t)$ for $t \in (t_k, t_{k+1})$. Since $y_1(t) < y_m(t) < z(t)$ for $t \in (t_k, t_{k+1})$, it follows from the condition (E) of Theorem 15.7 that there exists a subsequence $\{y_{m(j)}(t)\}$ such that $\{y_{m(j)}^{(i)}(t)\}$ converges uniformly to a solution $y(t)$ of (9.1) on compact subsets of $[a,b)$, where $y(t)$ satisfies

$$y^{(j)}(t_i) = x^{(j)}(t_i); \ 1 \leq i \leq r, \ i \neq k, \ 0 \leq j \leq \lambda(i)-1$$

$$y(t_k) = \gamma_0.$$

But, then $\gamma_0 \in S$ which is a contradiction.

To prove Theorem 18.8 when $\lambda(k)$ is an arbitrary positive integer we will use the following elementary lemma.

**Lemma 18.9**   Let $x(t)$ and $y(t)$ be in $C^{(p)}[\alpha,\beta]$, $p > 0$.   If $x^{(i)}(\alpha) = y^{(i)}(\alpha)$, $0 \le i \le p-1$, $x^{(p)}(\beta) \le y^{(p)}(\beta)$ and $x^{(p)}(t) < y^{(p)}(t)$ for $t$ e $[\alpha,\beta)$, then $x^{(i)}(t) < y^{(i)}(t)$ for $t$ e $(\alpha,\beta)$, $0 \le i \le p-1$.

**Theorem 18.10**   Let $1 \le k \le r$ and $s \le t_1 < \ldots < t_r < c_{\lambda(1)\ldots\lambda(r)}^{(s)}$, where $\lambda(k) \ge 2$.   If (9.1) satisfies condition (E) of Theorem 15.7 on $[a,b)$, then for any solution $x(t)$ of (9.1), $S(x;t_1^{\lambda(1)},\ldots,\bar{t}_k^{\lambda(k)},\ldots, t_r^{\lambda(r)})$ is an open interval.

**Proof.**   It follows from Corollary 18.7 that the set $S = S(x;t_1^{\lambda(1)},\ldots, \bar{t}_k^{\lambda(k)},\ldots,t_r^{\lambda(r)})$ is an open subset of R.   It suffices to establish that if $\gamma_0 = \sup\{\gamma : [x^{(\lambda(k)-1)}(t_k),\gamma] \subset S\}$ and $\gamma' > \gamma_0$, then $\gamma' \notin S$ and that, if $\lambda_0 = \inf\{\lambda : [\lambda,x^{(\lambda(k)-1)}(t_k)] \subset S\}$ and $\lambda' < \lambda_0$, then $\lambda' \notin S$. We will consider only the first case since the argument for the second case is similar.

Suppose $\gamma' > \gamma_0$ and $\gamma'$ e S.   Then, there must exist a solution $y(t)$ of (9.1) such that

$$y^{(j)}(t_i) = x^{(j)}(t_i); \quad 1 \le i \le r, \; i \ne k, \; 0 \le j \le \lambda(i)-1$$

$$y^{(j)}(t_k) = x^{(j)}(t_k), \quad 0 \le j \le \lambda(k)-2$$

$$y^{(\lambda(k)-1)}(t_k) = \gamma'.$$

Let $\{\gamma_m\}$ be a sequence of real numbers such that $x^{(\lambda(k)-1)}(t_k) < \gamma_m < \gamma_{m+1} < \gamma_0$ for each $m \ge 1$ and $\lim_{m\to\infty} \gamma_m = \gamma_0$.   Let $\{y_m(t)\}$ be the corresponding sequence of solutions of (9.1) such that

$$y_m^{(j)}(t_i) = x^{(j)}(t_i); \quad 1 \le i \le r, \; i \ne k, \; 0 \le j \le \lambda(i)-1$$

$$y_m^{(j)}(t_k) = x^{(j)}(t_k), \ 0 \le j \le \lambda(k)-2$$

$$y_m^{(\lambda(k)-1)}(t_k) = \gamma_m.$$

Assume that the sequence of functions $\{y_m^{(\lambda(k)-1)}(t)\}$ is uniformly bounded on $[t_k, t_k+\epsilon]$ for some $\epsilon > 0$. It then follows that $\{y_m(t)\}$ is uniformly bounded on $[t_k, t_k+\epsilon]$. But, then since (9.1) satisfies condition (E) of Theorem 15.7, there must exist a subsequence $\{y_{m(j)}(t)\}$ such that $\{y_{m(j)}(t)\}$ converges uniformly to a solution $z(t)$ of (9.1) on compact subsets of $[a,b)$, and consequently $\gamma_0 \in S$. This leads to a contradiction. Hence, $\{y_m^{(\lambda(k)-1)}(t)\}$ cannot be uniformly bounded on $[t_k, t_k+\epsilon]$ for any $\epsilon > 0$.

Since $\{y_m^{(\lambda(k)-1)}(t)\}$ cannot be uniformly bounded on $[t_k, t_k+\epsilon]$, for any $\epsilon > 0$ and $x^{(\lambda(k)-1)}(t_k) < y_m^{(\lambda(k)-1)}(t_k) < y^{(\lambda(k)-1)}(t_k)$, it follows by continuity that there exists a decreasing sequence $\{\delta_j\}$ such that $\delta_j > 0$, $\lim_{j\to\infty} \delta_j = 0$,

(18.2)
$$y_{m(j)}^{(\lambda(k)-1)}(t_k+\delta_j) = x^{(\lambda(k)-1)}(t_k+\delta_j)$$

or

(18.3)
$$y_{m(j)}^{(\lambda(k)-1)}(t_k+\delta_j) = y^{(\lambda(k)-1)}(t_k+\delta_j)$$

and

$$x^{(\lambda(k)-1)}(t) < y_{m(j)}^{(\lambda(k)-1)}(t) < y^{(\lambda(k)-1)}(t), \ t \in (t_k, t_k+\delta_j).$$

We will only consider the case (18.2) for $j \ge 1$.

By renumbering we can assume without loss of generality that

$$y_m^{(\lambda(k)-1)}(t_k+\delta_m) = x^{(\lambda(k)-1)}(t_k+\delta_m)$$

and

$$x^{(\lambda(k)-1)}(t) < y_m^{(\lambda(k)-1)}(t) < y^{(\lambda(k)-1)}(t), \ t \in (t_k, t_k+\delta_m).$$

By Lemma 18.9, $x^{(i)}(t) < y_m^{(i)}(t) < y^{(i)}(t)$; $t \in (t_k, t_k + \delta_m)$, $0 \le i \le \lambda(k) - 1$.
By the continuity of $x^{(i)}(t)$ and $y^{(i)}(t)$, we have $\lim\limits_{m \to \infty} y_m^{(i)}(t_k + \delta_m) =$
$x^{(i)}(t_k)$, $0 \le i \le \lambda(k) - 2$. But, then by Corollary 18.7 it follows that
$\lim\limits_{m \to \infty} y_m(t) = x(t)$ uniformly on compact subintervals of $[a,b]$ which leads
to a contradiction.

Theorem 18.11   Let $3 \le k \le p+1 \le n$ and $s \le t_1 < \ldots < t_{p+1} < c_{n-p,1,\ldots,1}(s)$.
If (9.1) satisfies condition (E) of Theorem 15.7 on $[a,b]$, then for any
solution $x(t)$ of (9.1)

$$S(x; t_1^{n-p}, t_2, \ldots, \bar{t}_k, \ldots, t_{p+1}) = S(x; t_1^{n-p+1}, t_3, \ldots, \bar{t}_k, \ldots, t_{p+1}).$$

Proof. By Theorem 18.8 and the fact that $c_{n-p,1,\ldots,1}(s) \le c_{n-p+1,1,\ldots,1}(s)$
(Theorem 18.4, part (i)), the sets $S_1 = S(x; t_1^{n-p}, t_2, \ldots, \bar{t}_k, \ldots, t_{p+1})$ and
$S_2 = S(x; t_1^{n-p+1}, t_3, \ldots, \bar{t}_k, \ldots, t_{p+1})$ are open intervals about $x(t_k)$. Let
$S_1 = (\lambda_1, \gamma_1)$ and $S_2 = (\lambda_2, \gamma_2)$. We will show that $\gamma_1 = \gamma_2$. The case
$\lambda_1 = \lambda_2$ follows from a similar argument.

Suppose $\gamma_1 < \gamma_2$. Let $z(t)$ be the solution of (9.1) such that

$$z^{(j)}(t_1) = x^{(j)}(t_1), \quad 0 \le j \le n-p$$

$$z(t_i) = x(t_i); \quad 3 \le i \le p+1, \quad i \ne k$$

$$z(t_k) = \frac{\gamma_1 + \gamma_2}{2}.$$

Let $0 < \delta < \min\{t_{k+1} - t_k, \ t_k - t_{k-1}\}$ (where $t_{p+2} = c_{n-p,1,\ldots,1}(s)$ in
case $k = p+1$). For $m$ an integer and $m > 1/(\gamma_1 - \gamma_2)$, there exists a unique solution $y_m(t)$ of (9.1) such that

$$y_m^{(j)}(t_1) = x^{(j)}(t_1), \ 0 \le j \le n-p-1$$

$$y_m(t_i) = x(t_i); \ 2 \le i \le p+1, \ i \ne k$$

$$y_m(t_k) = \gamma_1 - \frac{1}{m} \ .$$

Note that $\{y_m(t)\}$ is a strictly increasing sequence of functions on $(t_{k-1}, t_{k+1})$. As in the proof of Theorem 18.8, $\{y_m(t)\}$ cannot be uniformly bounded on any compact subinterval of $[a,b)$, hence for m sufficiently large there must exist $s_1 \in (t_{k-1}, t_k)$ and $s_2 \in (t_k, t_{k+1})$ such that $y_m(s_1) = z(s_1)$ and $y_m(s_2) = z(s_2)$. This is not possible, hence $\gamma_1 \ge \gamma_2$. A similar argument shows that $\gamma_1 \le \gamma_2$. This completes the proof of our theorem.

**Theorem 18.12** Let $2 \le k \le p \le n-1$ and $s \le t_1 < \ldots < t_{p+1} < c_{n-p,1,\ldots,1}(s)$. If (9.1) satisfies condition (E) of Theorem 15.7 on $[a,b]$, then for any solution $x(t)$ of (9.1)

$$S(x;t_1^{n-p},t_2,\ldots,\bar{t}_k,\ldots,t_{p+1}) = \begin{cases} S(x;t_1^{n-p+1},t_2,\ldots,\bar{t}_k,t_{k+2},\ldots,t_{p+1}), \\ \qquad\qquad \text{if } 2 \le k \le p-1 \\ \\ S(x;t_1^{n-p+1},t_2,\ldots,\bar{t}_p), \quad \text{if } k = p. \end{cases}$$

**Proof.** The proof is similar to that of Theorem 18.11.

**Theorem 18.13** Let $1 \le p \le n$ and $s \le t_1 < \ldots < t_{p+1} < c_{n-p,1,\ldots,1}(s)$. Assume that (9.1) satisfies condition (E) of Theorem 15.7 on $[a,b]$. If $S(y;t_1^{n-1},\bar{t}_k) = R$, for $2 \le k \le p+1$ and for every solution $y(t)$ of (9.1) satisfying $y^{(j)}(t_1) = A_{j+1,1}$, $0 \le j \le n-p-1$ then there exists a unique solution $x(t)$ of (9.1) such that $x^{(j)}(t_1) = A_{j+1,1}$, $0 \le j \le n-p-1$, $x(t_i) = A_{1,i}$, $2 \le i \le p+1$.

**Proof.** In view of Theorem 18.4 part (i), Theorems 18.11 and 18.12 imply that

$$S(y;t_1^{n-p},t_2,\ldots,\bar{t}_k,\ldots,t_{p+1}) = S(y;t_1^{n-p+1},t_3,\ldots,\bar{t}_k,\ldots,t_{p+1})$$

$$\vdots$$

$$= S(y;t_1^{n-p+k-2},\bar{t}_k,\ldots,t_{p+1})$$

$$= S(y;t_1^{n-p+k-1},\bar{t}_k,t_{k+2},\ldots,t_{p+1})$$

$$\vdots$$

$$= S(y;t_1^{n-1},\bar{t}_k) = R$$

for $2 \leq k \leq p+1$ and for every solution $y(t)$ of (9.1) satisfying $y^{(j)}(t_1) = A_{j+1,1}$, $0 \leq j \leq n-p-1$.

Let $P(q)$, $2 \leq q \leq p$ be the proposition : There exists a solution $x_q(t)$ of (9.1) such that

$$x_q^{(j)}(t_1) = A_{j+1,1}, \ 0 \leq j \leq n-p-1$$

$$x_q(t_i) = A_{1,i}, \ 2 \leq i \leq q$$

$$x_q(t_i) = x_{q-1}(t_i), \ q+1 \leq i \leq p+1$$

and

$$S(x_q;t_1^{n-p},t_2,\ldots,\bar{t}_{q+1},\ldots,t_{p+1}) = R.$$

We will show by induction on $q$ that $P(q)$ is true for $q = 2,\ldots,p$.

Assume $q = 2$. Let $x_1(t)$ be a solution of (9.1) which satisfies $x_1^{(j)}(t_1) = A_{j+1,1}$, $0 \leq j \leq n-p-1$. Since $S(x_1;t_1^{n-p},\bar{t}_2,\ldots,t_{p+1}) = R$, there exists a solution $x_2(t)$ of (9.1) such that

$$x_2^{(j)}(t_1) = x_1^{(j)}(t_1) = A_{j+1,1}, \ 0 \leq j \leq n-p-1$$

$$x_2(t_2) = A_{1,2}$$

$$x_2(t_i) = x_1(t_i), \quad 3 \le i \le p+1.$$

Since $x_2^{(j)}(t_1) = A_{j+1,1}$, $0 \le j \le n-p-1$, $S(x_2;t_1^{n-p},t_2,\bar{t}_3,\ldots,t_{p+1}) = R$.

Assume that $2 < q \le p$ and that $P(k)$ is true for $2 \le k < q$. Since $S(x_{q-1};t_1^{n-p},\ldots,\bar{t}_q,\ldots,t_{p+1}) = R$, there exists a solution $x_q(t)$ of (9.1) which satisfies $P(q)$. This completes the induction. Finally, since $S(x_p;t_1^{n-p},\ldots,\bar{t}_{p+1}) = R$, there exists a solution $x_{p+1}(t)$ of (9.1) satisfying

$$x_{p+1}^{(j)}(t_1) = A_{j+1,1}, \quad 0 \le j \le n-p-1$$

$$x_{p+1}(t_i) = A_{1,i}, \quad 2 \le i \le p+1.$$

**Theorem 18.14**   Let $2 \le k \le r$ and $s \le t_1 < \ldots < t_r < c_{\lambda(1)\ldots\lambda(r)}(s)$, where $\lambda(k) = 1$. Assume that (9.1) satisfies condition (E) of Theorem 15.7 on $[a,b)$. Further, we assume that every BVP : (9.1) together with only $(n-1)$ boundary conditions

(18.4)
$$x^{(j)}(t_i) = A_{j+1,i}; \quad 1 \le i \le r, \ i \ne k-1, \ 0 \le j \le \lambda(i)-1$$
$$x^{(j)}(t_{k-1}) = A_{j+1,k-1}, \quad 0 \le j \le \lambda(k-1)-2$$

has a solution. If $c_{\lambda(1),\ldots,\lambda(k-1)-1,1,\lambda(k),\ldots,\lambda(r)}(s) \ge c_{\lambda(1)\ldots\lambda(r)}(s)$, then $S(x;t_1^{\lambda(1)},\ldots,\bar{t}_k,\ldots,t_r^{\lambda(r)}) = R$ for every solution $x(t)$ of (9.1).

**Proof.**  It follows from Theorem 18.8 that $S(x;t_1^{\lambda(1)},\ldots,\bar{t}_k,\ldots,t_r^{\lambda(r)})$ is an open interval. Let $S = (\lambda,\gamma)$. We will show that $\gamma = \infty$. The proof that $\lambda = -\infty$ is similar. Assume that $\gamma < \infty$. Let $y(t)$ be a solution of (9.1) such that

$$y^{(j)}(t_i) = x^{(j)}(t_i); \quad 1 \le i \le r, \ i \ne k-1,k, 0 \le j \le \lambda(i)-1$$

$$y^{(j)}(t_{k-1}) = x^{(j)}(t_{k-1}), \ 0 \le j \le \lambda(k-1)-2$$

$$y(t_k) = \mu$$

where $\mu \ge \gamma$. Let $\gamma_m \in [x(t_k),\gamma)$ be such that $x(t_k) < \gamma_m < \gamma_{m+1}$ for each m, and $\lim\limits_{m \to \infty} \gamma_m = \gamma$. Let $\{y_m(t)\}$ be the corresponding sequence of solutions of (9.1) such that

$$y_m^{(j)}(t_i) = y^{(j)}(t_i); \ 1 \le i \le r, \ i \ne k, \ 0 \le j \le \lambda(i)-1$$

$$y_m(t_k) = \gamma_m.$$

It follows that $x(t) < y_m(t) < y_{m+1}(t)$ for all $t \in (t_{k-1},t_{k+1})$ (where $t_{k+1} = c_{\lambda(1)\ldots\lambda(r)}(s)$ in case $k = r$), and that $\{y_m(t)\}$ cannot be uniformly bounded on $[t_{k-1},t_{k+1}]$. For m sufficiently large there must be a $s_1 \in (t_{k-1},t_k)$ and a $s_2 \in (t_k,t_{k+1})$ such that $y_m(s_1) = y(s_1)$, and $y_m(s_2) = y(s_2)$, which is a contradiction.

**Theorem 18.15**  Let $s \le t_1 < t_2 < c_{pq}(s)$. Assume that (9.1) satisfies condition (E) of Theorem 15.7 on $[a,b]$. Further, we assume that every BVP : (9.1) together with only $(n-1)$ boundary conditions

$$x^{(j)}(t_1) = A_{j+1,1}, \ 0 \le j \le p-2$$

$$x^{(j)}(t_2) = A_{j+1,2}, \ 0 \le j \le q-1$$

has a solution. If $c_{pq}(s) \le \min\{c_{p,1,q-1}(s),c_{p,q-1,1}(s),c_{p-1,1,q-1,1}(s)\}$, then $S(x;t_1^p,t_2^q) = R$ for every solution x(t) of (9.1).

**Proof.**  It follows from Theorem 18.10 that $S(x;t_1^p,t_2^q)$ is an open interval, say, $S = (\lambda,\gamma)$. We will show that $\gamma = \infty$. Let $\gamma_m$ be such that

$x^{(q-1)}(t_2) < \gamma_m < \gamma_{m+1}$ for each m and $\lim\limits_{m\to\infty} \gamma_m = \gamma$. Let $\{y_m(t)\}$ be the corresponding sequence of solutions of (9.1) such that for each m

$$y_m^{(j)}(t_1) = x^{(j)}(t_1), \ 0 \le j \le p-1$$

$$y_m^{(j)}(t_2) = x^{(j)}(t_2), \ 0 \le j \le q-2$$

$$y_m^{(q-1)}(t_2) = \gamma_m.$$

Let y(t) be a solution of (9.1) such that

$$y^{(j)}(t_1) = x^{(j)}(t_1), \ 0 \le j \le p-2$$

$$y^{(j)}(t_2) = x^{(j)}(t_2), \ 0 \le j \le q-2$$

$$y^{(q-1)}(t_2) = \mu$$

where $\mu > \gamma$. We will assume that q is odd. The case where q is even is similar. Note that $y_{m+1}(t) - y_m(t)$ and $y_m(t) - x(t)$ have zeros of exact order q-1 at $t_2$ and $x^{(q-1)}(t_2) < y_m^{(q-1)}(t_2) < y_{m+1}^{(q-1)}(t_2)$. Since $c_{pq}(s) \le c_{p,1,q-1}(s)$, we have $x(t) < y_m(t) < y_{m+1}(t)$ for $t \in (t_1,t_2)$ and since $c_{pq}(s) \le c_{p,q-1,1}(s)$, we have $x(t) < y_m(t) < y_{m+1}(t)$ for $t \in (t_2,c_{pq}(s))$. Also, since $y(t) - y_m(t)$ and $y(t) - x(t)$ have zeros of exact order q-1 at $t_2$ and $x^{(q-1)}(t_2) < y_m^{(q-1)}(t_2) < y^{(q-1)}(t_2)$, we have $x(t) < y_m(t) < y(t)$ in a deleted neighborhood of $t_2$. The sequence of solutions $\{y_m(t)\}$ cannot be uniformly bounded on compact subsets of $[t_1,c_{pq}(s))$, hence for m sufficiently large there exist $s_1 \in (t_1,t_2)$ and $s_2 \in (t_2,c_{pq}(s))$ such that $y(s_1) = y_m(s_1)$ and $y(s_2) = y_m(s_2)$, which is a contradiction.

**Theorem 18.16**   Let $s \leq t_1 < \ldots < t_r < c_{\lambda(1)\ldots\lambda(r)}(s)$, where $\lambda(k) \geq 2$.

Assume that (9.1) satisfies condition (E) of Theorem 15.7 on $[a,b]$.

Further, we assume that every BVP : (9.1) together with only $(n-1)$

boundary conditions (18.4) has a solution.  If $c_{\lambda(1)\ldots\lambda(r)}(s) \leq$

$\min\{c_{\lambda(1),\ldots,\lambda(k-1),1,\lambda(k)-1,\ldots,\lambda(r)}(s), c_{\lambda(1),\ldots,\lambda(k-1),\lambda(k)-1,1,\ldots,}$

$\lambda(r)^{(s)}, c_{\lambda(1),\ldots,\lambda(k-1)-1,1,\lambda(k)-1,1,\ldots,\lambda(r)}(s)\}$, then $S(x; t_1^{\lambda(1)}, \ldots,$

$\bar{t}_k^{\lambda(k)}, \ldots, t_r^{\lambda(r)}) = R$.

**Proof.**  The proof is similar to that of Theorem 18.15.

The final result in this section provides a relation between the
first conjugate point for (9.1) and the first conjugate points of the
corresponding variational equations is due to Peterson [6] and Spencer
[13].

**Theorem 18.17**   Assume that the functions $f_i$, $0 \leq i \leq q$ where

$$f_i(t,x,\ldots,x^{(q)}) = \frac{\partial f}{\partial x^{(i)}}(t,x,\ldots,x^{(q)})$$

are continuous on $[a,b] \times R^{q+1}$, and the differential equation (9.1)
satisfies condition (E) of Theorem 15.7 on $[a,b]$.  Then,

$$\eta_1(s) = \inf_{(x_0(t))} \eta_1(s, x_0(t)).$$

## COMMENTS AND BIBLIOGRAPHY

Theorems 18.2-18.5 are adapted from [12] and are analogous to
several results proved in [4,5] for linear differential equations.
Theorem 18.5 is in fact an nth order analogue of Theorem 17.8.  The
"uniqueness implies existence" type results in this section are due to

234

Sukup [14]. Several other related results are available in [8,10].
More so, recently somewhat similar ideas have been used to study right
focal point BVPs [2,7,9].

1.  Azbelev, N. and Tsalyuk, Z. "On the question of the distribution
    of the zeros of solutions of a third order linear differential
    equation", Mat. Sb. (N.S.) 51, 475-486 (1960).
2.  Henderson, J. "Right focal point boundary value problems for
    ordinary differential equations and variational equations", J.
    Math. Anal. Appl. to appear.
3.  Peterson, A. C. "Boundary value functions for nonlinear differen-
    tial equations",Notices 20, A-119 (1973).
4.  Peterson, A. C. "On the ordering of multipoint boundary value
    functions", Canad. Math. Bull. 13, 507-513 (1970).
5.  Peterson, A. C. "On a relation between a theorem of Hartman and
    a theorem of Sherman", Canad. Math. Bull. 16, 275-281 (1973).
6.  Peterson, A. C. "An expression for the first conjugate point for
    an nth order nonlinear differential equation", Proc. Amer. Math.
    Soc. 61, 300-304 (1976).
7.  Peterson, A. C. "Existence-uniqueness for focal-point boundary
    value problems", SIAM J. Math. Anal. 12, 173-185 (1981).
8.  Peterson, A. C. "Existence-uniqueness for ordinary differential
    equations", J. Math. Anal. Appl. 64, 166-172 (1978).
9.  Peterson, A. C. "A disfocality function for a nonlinear ordinary
    differential equation", Rocky Mountain J. Math. 12, 741-752 (1982).
10. Peterson, A. C, and Sukup, D. V. "On the first conjugate point
    function for nonlinear differential equations", Canad. Math. Bull.
    18, 577-585 (1975).
11. Sherman, T. "Properties of solutions of nth order linear differen-
    tial equations", Pacific J. Math. 15, 1045-1060 (1965).
12. Spencer, J. D. "Boundary value functions for nonlinear differen-
    tial equations", J. Diff. Equs. 19, 1-20 (1975).
13. Spencer, J. D. "Relations between boundary value functions for a
    nonlinear differential equation and its variational equations",
    Canad. Math. Bull. 18, 269-276 (1975).
14. Sukup, D. V. "On the existence of solutions to multipoint boundary
    value problems", Rocky Mountain J. Math. 6, 357-375 (1976).

## 19. TOPOLOGICAL METHODS

A number of authors have employed various topological principles to study second order BVPs. For example, in [1-5,13,16] variations and refinements of Wazewski's topological method have been obtained for the first order differential system u' = F(t,u) and subsequently these are used to prove the existence and continuous dependence of solutions on boundary data for a class of second order BVPs more general than (7.1), (7.2). In [6] Wazewski's topological method together with the compactness properties of solution funnels have been used to prove the existence of solutions. In [15] Kelley has employed a method similar to Wazewski's to study the existence of the solutions of the BVP (9.1), (2.7) with q = n-1 and p = n-2. In this section, we shall use a newly developed topological transversality method due to Granas [10,11] which is a generalization of the continuation theorem of Leray and Schauder found in [9], to study the existence of solutions of a more general class of BVP

$$(19.1) \qquad L[x] = x^{(n)} + \sum_{i=1}^{n} p_i(t)x^{(n-i)} = f(t,\underline{x})$$

$$(19.2) \qquad U[x] = \ell$$

where $p_i(t)$, $1 \le i \le n$ are continuous on $I = [a,b]$, f is continuous on $I \times R^{q+1}$, $U : C^{(n-1)}(I) \to R^n$ is a continuous linear operator and $\ell$ is a given vector in $R^n$.

To establish the existence of solutions of the BVP (19.1), (19.2) we need to consider solutions of an associated family of differential equations

$$(19.3) \qquad L[x] = g(t,\underline{x},\lambda)$$

satisfying (19.2). In (19.3) the function $g : I \times R^{q+1} \times [0,1] \to R$ is continuous, $g(t,\underline{x},0) \equiv 0$ and $g(t,\underline{x},1) \equiv f(t,\underline{x})$. An obvious choice of g is $\lambda f$.

The following definitions and Lemmas 19.1 and 19.2 are due to Granas [10].

**Definitions 19.1** Assume all topological spaces are Hausdroff. Let Y be a topological space, $A \subset X \subset Y$ and A closed in X.

(i)    A continuous mapping $f : X \to Y$ is compact if $\overline{f(X)}$ is compact.

(ii)    $h : [0,1] \times X \to Y$ is a compact homotopy if h is a homotopy and if for each $\lambda \in [0,1]$, $h|\lambda \times X \equiv h_\lambda$ is compact.

(iii)    $f : X \to Y$ is admissible with respect to A if f is compact and $f|A$ is fixed point free. Let $M_A(X,Y)$ denote the class of admissible mappings with respect to A.

(iv)    $f \in M_A(X,Y)$ is inessential if there exists $g \in M_A(X,Y)$ such that $f|A = g|A$ and g is fixed point free on X. Otherwise, $f \in M_A(X,Y)$ is essential.

(v)    A compact homotopy $h : [0,1] \times X \to Y$ is admissible if for each $\lambda \in [0,1]$, $h_\lambda$ is admissible. Two mappings $f,g \in M_A(X,Y)$ are homotopic in $M_A(X,Y)$, $f \sim g$, if there exists an admissible homotopy $h : [0,1] \times X \to Y$ such that $h_0 = g$ and $h_1 = f$.

(vi)    F* denotes the class of topological spaces which has the fixed point property for compact maps. We remark that a closed convex subspace of a Banach space is a F* space by the Schauder fixed point theorem.

**Lemma 19.1** Let Y be a connected space belonging to F*, let $X \subset Y$ be closed, and let $A = \partial X$. If $f : X \to Y$ is a constant mapping, i.e., $f(x) = p$ for all $x \in X$, and $p \in X \backslash A$, then f is essential.

**Lemma 19.2** Let Y be a convex topological space and let A and X be as in Lemma 19.1. Assume $f \sim g$ in $M_A(X,Y)$. Then, f is essential if and only if g is essential.

Let $(C(I), |\cdot|_0)$ be the Banach space of continuous functions on $I$ with supremum norm and let $(C^{(k)}(I), |\cdot|_k)$ be the Banach space of $k$ times continuously differentiable functions $x$ with norm

$$|x|_k = \max\{|x|_0, |x'|_0, \ldots, |x^{(k)}|_0\}.$$

We shall now employ Granas' topological transversality method to prove a general theorem concerning the existence of the BVP (19.1), (19.2).

**Theorem 19.3**   Assume that $x \equiv 0$ is the unique solution of the BVP $L[x] = 0$, $U[x] = 0$ and there exists $C > 0$ such that $|x|_{n-1} < C$ for all solutions $x$ of the BVP (19.3), (19.2) for all $0 \leq \lambda \leq 1$.  Then, the BVP (19.1), (19.2) has at least one solution.

**Proof.**   Let $Y = C^{(n-1)}(I)$, $X = \{x \in C^{(n-1)}(I) : |x|_{n-1} \leq C\}$ and $A = \partial X$. Since $x \equiv 0$ is the unique solution of the BVP $L[x] = 0$, $U[x] = 0$ there exists a Green's function $G(t,s)$ for the BVP $L[x] = 0$, $U[x] = 0$.  Define $h : [0,1] \times X \to Y$ as follows

$$h(\lambda, x)(t) = P_\ell(t) + \int_a^b G(t,s) g(s, \underline{x}(s), \lambda) ds$$

where $P_\ell(t)$ is the unique solution of the BVP $L[x] = 0$, $U[x] = \ell$.  For each $\lambda$, $0 \leq \lambda \leq 1$, $h_\lambda : X \to Y$ can be shown to be a compact map by an application of the Arzela-Ascoli theorem and since all solutions $x$ of (19.3), (19.2) satisfy $|x|_{n-1} < C$, each $h_\lambda$ is admissible.  Thus, $h$ is an admissible homotopy and $h_0 \sim h_1$.

Now, $h_0 \equiv P_\ell$ and $P_\ell \in X/A$ since $|P_\ell|_{n-1} < C$.  Thus, by Lemma 19.1 $h_0$ is essential.  It follows from Lemma 19.2 that $h_1$ is essential.  Since the BVPs (19.1), (19.2) and (19.3), (19.2) with $\lambda = 1$ are equivalent, the BVP (19.1), (19.2) has at least one solution.

The applicability of Theorem 19.3 depends upon the existence of an a priori $|\cdot|_{n-1}$ norm bound for solutions of the family of BVPs (19.3), (19.2) which is independent of $\lambda$. In the next result, we shall provide sufficient conditions under which such bounds exist for the BVP (9.1), (3.8).

**Theorem 19.4**  Assume that there exists a positive real valued function $\phi(y_0,\ldots,y_q)$ defined for $y_i \geq 0$, $0 \leq i \leq q$ which is nondecreasing in each variable and such that

$$|f(t,x_0,\ldots,x_q)| \leq \phi(|x_0|,\ldots,|x_q|)$$

for all $(t,\underline{x})$ e $I \times R^{q+1}$.  If

(19.4) $$\sum_{i=0}^{q} \frac{y_i}{\phi(y_0,\ldots,y_q)} \longrightarrow \infty \text{ as } \sum_{i=0}^{q} y_i \longrightarrow \infty,$$

then the BVP (9.1), (3.8) has a solution on $[a,b]$.

**Proof.**  Obviously, $x \equiv 0$ is the only solution of the BVP (3.7), (3.8). Thus, by Theorem 19.3, the proof is complete if we exhibit $C > 0$ such that $|x|_{n-1} < C$ for all solutions $x$ of the family of differential equations

(19.5) $$x^{(n)} = \lambda f(t,\underline{x})$$

satisfying (3.8).

Let $x$ be a solution of (19.5), (3.8) then from Theorem 8.1, we have

$$|x^{(k)}(t)| \leq \frac{1}{(n-k)!} (b-a)^{n-k} \max_{a \leq t \leq b} |x^{(n)}(t)|$$

$$= \frac{1}{(n-k)!} (b-a)^{n-k} \max_{a \leq t \leq b} |\lambda f(t,\underline{x}(t))|$$

$$\leq \frac{1}{(n-k)!} (b-a)^{n-k} \max_{a \leq t \leq b} \phi(|x(t)|,\ldots,|x^{(q)}(t)|)$$

$$\leq \frac{1}{(n-k)!} (b-a)^{n-k} \phi(|x|_0,\ldots,|x^{(q)}|_0)$$

and hence

(19.6) $$|x^{(k)}|_0 \leq K \phi (|x|_0,\ldots,|x^{(q)}|_0), \quad 0 \leq k \leq n-1$$

where $K = \max_{0 \leq k \leq n-1} \frac{1}{(n-k)!} (b-a)^{n-k}$.

Thus, for all solutions x of (19.5), (3.8) we find

$$\sum_{k=0}^{q} \frac{|x^{(k)}|_0}{\phi(|x|_0,\ldots,|x^{(q)}|_0)} \leq K(q+1).$$

By (19.4), this implies that there exists $K_1 > 0$, independent of $\lambda$, such that $|x^{(i)}|_0 < K_1$, $0 \leq i \leq q$ for all solutions x of the family of BVPs (19.5), (3.8). Hence, from (19.6) it follows that $|x^{(n-1)}|_0 < K\phi(K_1,\ldots,K_1) = C$.

Remark 19.1   As in Theorem 9.1 and Corollary 9.9 the Schauder fixed point theorem can be employed to obtain Theorem 19.4.

In our final application of Theorem 19.3, we consider the BVP (9.1), (19.2). We shall impose on f a Nagumo-like estimate as in Jackson [14], in order to allow a faster growth rate on f than that allowed by (19.4). For this, we need

Lemma 19.5 [12]   Let $x \in C^{(n)}(I)$. Then, for each integer k, $0 < k < n$

$$|x^{(k)}|_0 \leq F_{n,k} |x|_0^{1-k/n} M_n^{k/n}$$

where $F_{n,k} = 4e^{2k}n^k k^{-k}$ and $M_n = \max\{|x^{(n)}|_0, \; 2^n n! \, |x|_0 (b-a)^{-n}\}$.

**Theorem 19.6** Given any $M > 0$, assume that there exists a positive real valued function $\phi(y_1, \ldots, y_q)$ defined for $y_i \geq 0$, $1 \leq i \leq q$ which is nondecreasing in each variable and such that

$$|f(t, x_0, \ldots, x_q)| \leq \phi(|x_1|, \ldots, |x_q|)$$

for all $(t, \underline{x}) \in I \times [-M, M] \times R^q$, and

$$(19.7) \qquad \sum_{i=1}^{q} \frac{y_i^{n/i}}{\phi(y_1, \ldots, y_q)} \longrightarrow \infty \text{ as } \sum_{i=1}^{q} y_i^{n/i} \longrightarrow \infty.$$

If there exists $N > 0$ such that $|x|_0 < N$ for all solutions x of the associated family of BVPs (19.5), (19.2) for all $0 \leq \lambda \leq 1$, then the BVP (9.1), (19.2) has a solution on $[a,b]$.

**Proof.** Let x be a solution of the BVP (19.5), (19.2) and assume that $|x|_0 < N$. By Lemma 19.5, there exist constants $C_1(N)$ and $C_2(N)$, independent of $\lambda$ and x, such that

$$|x^{(k)}|_0^{n/k} \leq C_1(N) \leq \frac{C_1(N)}{\phi(0, \ldots, 0)} \phi(|x'|_0, \ldots, |x^{(q)}|_0)$$

or

$$|x^{(k)}|_0^{n/k} \leq C_2(N) |x^{(n)}|_0 \leq C_2(N) \phi(|x'|_0, \ldots, |x^{(q)}|_0), \; 1 \leq k \leq n-1.$$

Consequently, there exists a constant $C(N)$, independent of $\lambda$ and x, such that

$$\sum_{i=1}^{q} \frac{|x^{(i)}|_0^{n/i}}{\phi(|x'|_0, \ldots, |x^{(q)}|_0)} \leq C(N).$$

By (19.7), it follows as in the proof of Theorem 19.4 that there exists $C > 0$, independent of $\lambda$, such that $|x|_{n-1} < C$. This completes the proof.

## COMMENTS AND BIBLIOGRAPHY

Theorem 19.3 uses technique due to Granas [9,10] and is adapted
from [7]. Theorems 19.4 and 19.6 are modelled after the work of Eloe
and Henderson [7]. Let $U : [\frac{1}{2},1] \times C^{(n-1)}(I) \to R^n$ be continuous and for
each $\frac{1}{2} \le \lambda \le 1$, $U(\lambda)$ is a linear operator. In [8] using the same technique the existence of solutions of the differential equation (19.1)
satisfying $U(\lambda)x = \ell$ is proved.

1.    Bebernes, J. W. and Fraker, R. "A priori bounds for boundary
      sets", Proc. Amer. Math. Soc. 29, 313-318 (1971).
2.    Bebernes, J. W. and Kelley, W. G. "Some boundary value problems
      for generalized differential equations", SIAM J. Appl. Math. 25,
      16-23 (1973).
3.    Bebernes, J. W. and Schuur, J. D. "Investigations in the topologi-
      cal method of Wazewski", Atti della Accademia Nazionale dei Lincei.
      Classe di Scienze Fisiche, Matematiche e Naturali 49, 39-42 (1970).
4.    Bebernes, J. W. and Schuur, J. D. "The Wazewski topological method
      for contingent equations", Ann. Mat. Pura Appl. 87, 271-280 (1970).
5.    Bebernes, J. W. and Wilhelmsen, R. "A technique for solving two-
      dimensional boundary value problems", SIAM J. Appl. Math. 17,
      1060-1064 (1969).
6.    Bebernes, J. W. and Wilhelmsen, R. "A general boundary value tech-
      nique", J. Diff. Equs. 8, 404-415 (1970).
7.    Eloe, P. W. and Henderson, J. "Nonlinear boundary value problems
      and a priori bounds on solutions", SIAM J. Math. Anal. 15, 642-
      647 (1984).
8.    Eloe, P. W. and Henderson, J. "Families of boundary conditions for
      nonlinear ordinary differential equations", Nonlinear Analysis :
      Theory, Methods and Appl. 9, 631-638 (1985).
9.    Gaines, R. E. and Mawhin, J. L. Coincidence Degree, and Nonlinear
      Differential Equations, Lecture Notes in Mathematics 568, Springer-
      Verlag, New York, 1977.
10.   Granas, A. "Sur la méthode de continuité de Poincaré", C. R. Acad.
      Sci. Paris 282, 983-985 (1976).
11.   Granas, A. Guenther, R. B. and Lee, J. W. "Nonlinear boundary value
      problems for some classes of ordinary differential equations",
      Rocky Mountain J. Math. 10, 35-58 (1980).
12.   Gorny, A. "Contribution a l'étude des fonctions dérivables d'une
      variable réelle", Acta Math. 71, 317-358 (1939).
13.   Jackson, L. K. and Klaasen, G. "A variation of the topological
      method of Wazewski", SIAM J. Appl. Math. 20, 124-130 (1971).
14.   Jackson, L. K. "A Nagumo condition for ordinary differential equa-
      tions", Proc. Amer. Math. Soc. 57, 93-96 (1976).

242

15. Kelley, W. G. "Some existence theorems for nth-order boundary value problems", J. Diff. Equs. 18, 158-169 (1975).
16. Sedziwy, S. "Dependence of solutions on boundary data for a system of two ordinary differential equations", J. Diff. Equs. 9, 381-389 (1971).

## 20. BEST POSSIBLE RESULTS : CONTROL THEORY METHODS

In Section 10, we have used Contraction Mapping Principle and obtained the conclusion that a given BVP has a unique solution provided certain inequality over the length of the interval is satisfied. In general this inequality does not provide the best possible length estimates in the sense that unique solutions may exist on longer intervals. The results obtained in Sections 12 and 13 are some examples where the weight function technique or shooting methods indeed give best possible intervals. In this section, we shall apply techniques from optimal control theory to obtain some more best possible results.

Throughout, we shall assume that the function $f(t,\underline{x})$ is continuous on $(a,b) \times R^{q+1}$ and satisfies the Lipschitz condition (9.12). Thus, the differential equation (9.1) obviously satisfies the conditions (A), (B) and (C) of Theorem 15.5. Furthermore, from Theorem 9.12 all n point BVPs (9.1), (2.3) are uniquely solvable on small subintervals of $(a,b)$ and hence condition $(E_1)$ which is equivalent to condition (E) of Theorem 15.7 (see Remark 15.5) is satisfied. Thus, if the differential equation (9.1) also satisfies condition (D) of Theorem 15.5 (condition $(D_2)$ of Theorem 15.9) on a subinterval $(\alpha,\beta) \subset (a,b)$, then from Theorem 15.8 (Theorem 15.9) all r point (right $(m_1,\ldots,m_r)$ focal point) BVPs will have unique solutions on $(\alpha,\beta)$. Our aim is now to characterize in terms of the Lipschitz constants $L_j$, $0 \leq j \leq q$ the subintervals $(\alpha,\beta)$ of $(a,b)$ of maximal length on which condition (D) of Theorem 15.5 (condition $(D_2)$ of Theorem 15.9) is satisfied. For this, we begin with the necessary preliminaries so that the Pontryagin Maximum Principle [4, p.314] can be applied.

Let U be the set of all vector functions $u = (u_0(t),\ldots,u_q(t))$ such that the component functions $u_j(t)$ are Lebesgue measurable on $(a,b)$ and satisfy the inequalities $|u_j(t)| \leq L_j$, $0 \leq j \leq q$. Let I, J be nonempty subsets of $\{1,\ldots,n\}$ such that card (I) + card (J) = n, and let $I^c$, $J^c$ denote the respective complements of I, J in $\{1,\ldots,n\}$. For fixed such sets of integers I and J, we consider the BVPs

(20.1)
$$x^{(n)} = \sum_{j=0}^{q} u_j(t) x^{(j)}$$

(20.2)     $x^{(i-1)}(c) = 0$ for i e I

(20.3)     $x^{(i-1)}(d) = 0$ for i e J

where $a < c < d < b$ and $u = (u_0(t),\ldots,u_q(t))$ e U. If there exist
$a < c < d < b$ and u e U such that the corresponding problem (20.1) -
(20.3) has a nontrivial solution, then there is a time optimal solution,
i.e., there is a u* e U and $c \le c_1 < d_1 \le d$ such that the BVP

$$x^{(n)} = \sum_{j=0}^{q} u_j^*(t) x^{(j)}$$

$$x^{(i-1)}(c_1) = 0 \text{ for i e I}$$

$$x^{(i-1)}(d_1) = 0 \text{ for i e J}$$

has a nontrivial solution $x(t)$ and $d_1 - c_1$ is a minimum over all such
solutions. For this time optimal solution $x(t)$ let $z(t) = (x(t), x'(t),$
$\ldots, x^{(n-1)}(t))^T$, then $z(t)$ is a solution of the corresponding first
order system $z' = A[u*(t)]z$.

By the Pontryagin Maximum Principle, the adjoint system

(20.4)          $\psi' = -A^T[u*(t)]\psi$

has a solution $\psi(t) = (\psi_1(t),\ldots,\psi_n(t))^T$ such that

(20.5)     $\sum_{j=1}^{n} x^{(j)}(t)\psi_j(t) = \langle z'(t), \psi(t) \rangle$

$$= \max_{u e U} \{\langle A[u(t)]z(t), \psi(t) \rangle\}$$

for almost all t with $c_1 \le t \le d_1$, where $<\cdot,\cdot>$ is the inner product. Furthermore, $<z'(t),\psi(t)>$ is a nonnegative constant for almost all $c_1 \le t \le d_1$ and $\psi_j(c_1) = 0$ for $j \in I^c$ and $\psi_j(d_1) = 0$ for $j \in J^c$. Since

$$<A[u(t)]z(t),\psi(t)> = \sum_{j=1}^{n-1} x^{(j)}(t)\psi_j(t) + \psi_n(t) \sum_{j=0}^{q} u_j(t)x^{(j)}(t)$$

the maximum condition (20.5) can be written as

$$(20.6) \quad \psi_n(t) \sum_{j=0}^{q} u_j^*(t)x^{(j)}(t) = \max_{u \in U} \{\psi_n(t) \sum_{j=0}^{q} u_j(t)x^{(j)}(t)\}$$

from which it follows that if $\psi_n(t)$ has no zeros on $(c_1,d_1)$ and if $x(t) > 0$ on $(c_1,d_1)$, then (20.6) determines the optimal control $u^*(t)$.

In particular, if $x(t) > 0$ and $\psi_n(t) < 0$ on $(c_1,d_1)$, then (20.6) implies that the time optimal solution $x(t)$ is a solution of

$$(20.7) \qquad x^{(n)} = -[L_0 x + \sum_{j=1}^{q} L_j \, | \, x^{(j)} \, |]$$

on $[c_1,d_1]$. On the other hand, if $x(t) > 0$ and $\psi_n(t) > 0$ on $(c_1,d_1)$, then the time optimal solution $x(t)$ is a solution of

$$(20.8) \qquad x^{(n)} = L_0 x + \sum_{j=1}^{q} L_j \, | \, x^{(j)} \, |$$

on $[c_1,d_1]$.

We also note that if there is a $u \in U$ such that the problem (20.1)-(20.3) has a nontrivial solution, then the problem

$$(20.9) \qquad \psi' = -A^T[u(t)]\psi$$

$$(20.10) \qquad \psi_j(c) = 0 \text{ for } j \in I^c$$

(20.11)
$$\psi_j(d) = 0 \text{ for } j \in I^d$$

has a nontrivial solution, and conversely. Thus, the Maximum Principle associates with a time optimal solution of (20.1)-(20.3) a time optimal solution of (20.9)-(20.11).

**Definition 20.1** The boundary conditions (2.4) with $A_{j+1,i} = 0$; $1 \le i \le r$, $0 \le j \le k_i$ are called $(k_1+1,\dots,k_r+1)$ conjugate boundary conditions. If an equation (20.1) has no nontrivial solution satisfying these boundary conditions for any $a_1 < a_r$ in an interval, we say that the equation is $(k_1+1,\dots,k_r+1)$ disconjugate on that interval. Similarly, the boundary conditions (15.6) with $A_{ij} = 0$; $s_{j-1} \le i \le s_j-1$, $1 \le j \le r$ are called $(m_1,\dots,m_r)$ focal boundary conditions and an equation (20.1) will be called $(m_1,\dots,m_r)$ disfocal on an interval if it has no nontrivial solutions satisfying these conditions on that interval.

**Theorem 20.1** If there is a control vector $u \in U$ such that the corresponding equation (20.1) has a nontrivial solution satisfying $(n-1,1)$ conjugate boundary conditions on $(a,b)$ and if $x(t)$ is a time optimal solution satisfying $x^{(i-1)}(c) = 0$, $1 \le i \le n-1$, $x(d) = 0$ with $d - c$ a minimum, then $x(t)$ is a solution of (20.7) on $[c,d]$. If for all $u \in U$ and all $j$ with $k+1 \le j \le n-1$, $1 \le k \le n-1$ the corresponding equations (20.1) are $(j,n-j)$ disconjugate on $(a,b)$ and if there is an equation in the collection (20.1) which has a nontrivial solution satisfying $(k,n-k)$ conjugate boundary conditions on $(a,b)$, then a time optimal solution $x(t)$ with $x^{(i-1)}(c) = 0$, $1 \le i \le k$, $x^{(i-1)}(d) = 0$, $1 \le i \le n-k$ and $d - c$ a minimum is a solution of (20.7) on $[c,d]$ when $n - k$ is odd, and is a solution of (20.8) on $[c,d]$ when $n - k$ is even.

**Proof** Let $x(t)$ be a time optimal solution satisfying $(n-1,1)$ conjugate boundary conditions $x^{(i-1)}(c) = 0$, $1 \le i \le n-1$, $x(d) = 0$ with $d - c$ a minimum. Then because of the time optimality $x(t) \ne 0$ for $c < t < d$ and without loss of generality we can assume that $x(t) > 0$ on $(c,d)$. It follows that $x^{(n-1)}(c) > 0$. Since the solution $\psi(t)$ of the adjoint system associated with $x(t)$ by the Maximum Principle satisfies $\psi_n(c) = 0$,

$\psi_i(d) = 0$, $2 \leq i \leq n$ and is also time optimal, it follows that $\psi_n(t) \neq 0$ for $c < t < d$. Thus, $x(t)$ is a solution of either (20.7) or (20.8) on $[c,d]$ which means that in either case $x^{(n-1)}(t)$ is strictly monotone on $[c,d]$. We also have

$$\sum_{j=1}^{n} x^{(j)}(t)\psi_j(t) = x^{(n-1)}(c)\psi_{n-1}(c) = x'(d)\psi_1(d)$$

on $c \leq t \leq d$ with the constant value being nonnegative. If $x'(d) = 0$, then from Rolle's theorem $x^{(n-1)}(t)$ has two zeros on $(c,d)$ which contradicts the strict monotonicity of $x^{(n-1)}(t)$. Also, $\psi_1(d) \neq 0$ since $\psi(t)$ is a nontrivial solution of the adjoint system. Thus, $x'(d)\psi_1(d) \neq 0$ and $x^{(n-1)}(c)\psi_{n-1}(c) > 0$. Since $x^{(n-1)}(c) > 0$, we find that $\psi_{n-1}(c) > 0$. Using this, the adjoint system provides $\psi_n'(c) < 0$, and hence $\psi_n(t) < 0$ on $(c,d)$. Thus, $x(t)$ is a solution of (20.7) on $[c,d]$.

Now, assume that for each $u \in U$ the corresponding equation (20.1) is $(j,n-j)$ disconjugate on $(a,b)$ for each $j$ with $k+1 \leq j \leq n-1$, but assume that there is an equation (20.1) with a nontrivial solution satisfying $(k,n-k)$ conjugate boundary conditions. Let $x(t)$ be a time optimal such solution with $x^{(i-1)}(c) = 0$, $1 \leq i \leq k$, $x^{(i-1)}(d) = 0$, $1 \leq i \leq n-k$ and $d - c$ a minimum. Then, it follows from lemma 4 in [8] that $x(t) \neq 0$ for $c < t < d$ and again without loss of generality we can assume $x(t) > 0$ on $(c,d)$. Since all equations (20.1) with $u \in U$ are $(k+1,n-k-1)$ disconjugate on $(a,b)$, it follows that $x^{(k)}(c) \neq 0$, and hence $x^{(k)}(c) > 0$. Let $\psi(t)$ be the solution of the adjoint system associated with $x(t)$, then $\psi_n(t) \neq 0$ on $(c,d)$ follows from the corollary to theorem 6 in [2]. Hence, again $x(t)$ is a solution of either (20.7) or (20.8) on $[c,d]$. Now, as in the proof of the first part of the theorem, this implies $x^{(n-k)}(d) \neq 0$. Furthermore, because of the adjoint system having only the trivial solution satisfying $\psi_i(c) = 0$, $k+2 \leq i \leq n$, $\psi_i(d) = 0$, $n-k \leq i \leq n$ we conclude that $\psi_{n-k}(d) \neq 0$ for the solution of the adjoint system associated with $x(t)$. Thus, we have

$$\sum_{j=1}^{n} x^{(j)}(t)\psi_j(t) = x^{(k)}(c)\psi_k(c) = x^{(n-k)}(d)\psi_{n-k}(d) \neq 0$$

and from $x^{(k)}(c) > 0$, we conclude that $\psi_k(c) > 0$. An examination of the adjoint system leads to the conclusion that $\psi_j(c) = 0$, $k+1 \leq j \leq n$ and $\psi_k(c) > 0$ implies $\text{Sgn } \psi_n(t) = (-1)^{n-k}$ on $(c,d)$. Thus, $x(t)$ is a solution of (20.7) on $[c,d]$ if $n-k$ is odd and is a solution of (20.8) on $[c,d]$ if $n-k$ is even.

**Theorem 20.2** If there is a control $u \in U$ such that the corresponding equation (20.1) has a nontrivial solution satisfying $(n-1,1)$ focal boundary conditions on $(a,b)$ and if $x(t)$ is a time optimal solution with $x^{(i-1)}(c) = 0$, $1 \leq i \leq n-1$, $x^{(n-1)}(d) = 0$ and $d - c$ a minimum, then $x(t)$ is a solution of (20.7) on $[c,d]$. If for all $u \in U$ and all $j$ with $k+1 \leq j \leq n-1$ the corresponding equations (20.1) are $(j,n-j)$ disfocal on $(a,b)$ and if there is an equation in the collection (20.1) which has a nontrivial solution satisfying $(k,n-k)$ focal boundary conditions on $(a,b)$, then a time optimal solution $x(t)$ with $x^{(i-1)}(c) = 0$, $1 \leq i \leq k$, $x^{(i-1)}(d) = 0$, $k+1 \leq i \leq n$ and $d - c$ a minimum is a solution of (20.7) on $[c,d]$ if $n-k$ is odd and is a solution of (20.8) on $[c,d]$ if $n - k$ is even.

**Proof.** Let $x(t)$ be a time optimal solution satisfying $(n-1,1)$ focal boundary conditions $x^{(i-1)}(c) = 0$, $1 \leq i \leq n-1$, $x^{(n-1)}(d) = 0$ and $d - c$ a minimum. Then, the associated solution $\psi(t)$ of the adjoint system is a time optimal solution satisfying the conditions $\psi_n(c) = 0$, $\psi_j(d) = 0$ $1 \leq j \leq n-1$. It follows that $\psi_n(t) \neq 0$ on $(c,d]$. Since $x^{(n-1)}(c) \neq 0$, we can assume that $x^{(n-1)}(c) > 0$ and $x(t) > 0$ on $(c,d]$. Thus, $x(t)$ is a solution of either (20.7) or (20.8) on $[c,d]$ and it is of (20.7) follows as in Theorem 20.1.

Now, assume that for each $u \in U$ the corresponding equation (20.1) is $(j,n-j)$ disfocal on $(a,b)$ for each $k+1 \leq j \leq n-1$ but there is an equation (20.1) with a nontrivial solution satisfying $(k,n-k)$ focal boundary conditions on $(a,b)$. Let $x(t)$ be a time optimal such solution with $x^{(i-1)}(c) = 0$, $1 \leq i \leq k$, $x^{(i-1)}(d) = 0$, $k+1 \leq i \leq n$ and $d - c$ a minimum. For the equation (20.1) of which $x(t)$ is a solution let $d_0$

be the infimum of all s with $c < s < b$ such that there is a nontrivial solution $y(t)$ satisfying $y^{(i-1)}(c) = 0$, $1 \leq i \leq k$, $y^{(i-1)}(t_i) = 0$, $k+1 \leq i \leq n$ where $c \leq t_{k+1} \leq \cdots \leq t_n \leq s$. Then, $c < d_0 \leq d$. Now, following the proof of proposition 1 in [7], it is easy to prove that there is a nontrivial solution $y(t)$ such that for some m with $k \leq m \leq n-1$, $y^{(i-1)}(c) = 0$, $1 \leq i \leq m$, $y^{(i-1)}(d_0) = 0$, $m+1 \leq i \leq n$ and $y^{(i-1)}(t) \neq 0$ on $(c,d_0)$ for $1 \leq i \leq m$. It follows from the (j,n-j) disfocality for $k+1 \leq j \leq n-1$ that $m = k$. It then follows from the optimality of the solution $x(t)$ that $d_0 = d$ and then from the (k+1,n-k-1) disfocality that the above extremal solution $y(t)$ is a scalar multiple of $x(t)$. Hence, $x^{(i-1)}(t) \neq 0$ on $(c,d)$ for $1 \leq i \leq k$. It follows that $x(t) \neq 0$ on $(c,d]$ since otherwise the preceding assertion would be contradicted by Rolle's theorem. We can assume then that $x^{(k)}(c) > 0$ and $x(t) > 0$ on $(c,d]$.

Let $\psi(t)$ be the time optimal solution of the adjoint system associated with $x(t)$, then $\psi_i(c) = 0$, $k+1 \leq i \leq n$, $\psi_i(d) = 0$, $1 \leq i \leq k$. If we reverse the order of the components of $\psi(t)$, i.e., define the vector function $y(t) = (y_1(t), \ldots, y_n(t))$ by $y_j(t) = \psi_{n+1-j}(t)$ for $1 \leq j \leq n$, then $y(t)$ is a solution of a first order system of type considered in [1]. Furthermore, $y_i(c) = 0$, $1 \leq i \leq n-k$, $y_i(d) = 0$, $n-k+1 \leq i \leq n$ so that

$$y(t) = \sum_{j=1}^{n-k} c_j \, y^j(t)$$

where $y^j(t)$ is the solution of our modified first order system such that $y_i^j(d) = \delta_{ij}$, $1 \leq i, j \leq n$. If $c_{n-k} = 0$, then $y_i(d) = 0$ for $i = n-k$ and we conclude that $\psi(t)$ is a nontrivial solution of the adjoint system with $\psi_i(c) = 0$, $k+2 \leq i \leq n$, $\psi_i(d) = 0$, $1 \leq i \leq k+1$. This in turn would imply the existence of a nontrivial solution $z(t)$ of our time optimal equation from the collection (20.1) with $z^{(i-1)}(c) = 0$, $1 \leq i \leq k+1$, $z^{(i-1)}(d) = 0$, $k+2 \leq i \leq n$. This contradicts the (k+1,n-k-1) disfocality, and hence $c_{n-k} \neq 0$.

If $W(y^1, \ldots, y^j)(t)$ is the determinant in which the ith row is

$(y_i^1(t),\ldots,y_i^j(t))$ for $1 \le i \le j$, then it follows that $W(y^1,\ldots,y^j)(t) \ne 0$ on $(a,b)$ for $1 \le j \le n-k-1$ because of the $(j,n-j)$ disfocality on $(a,b)$ for $k+1 \le j \le n-1$. With these conditions satisfied, if $y_1(t) = 0$ at some point in $(c,d]$, we can apply theorem 2.1 of [1] successively to reach the conclusion that $W(y^1,\ldots,y^{n-k})(t_0) = 0$ for some $c < t_0 < d$. This contradicts the time optimality of $\psi(t)$ and we conclude that $y_1(t) = \psi_n(t) \ne 0$ on $(c,d]$. Thus, the time optimal solution $x(t)$ of our $(k,n-k)$ focal BVP is either a solution of (20.7) on $[c,d]$ or a solution of (20.8) on $[c,d]$. In either case, from the form of equations (20.7) and (20.8) we see that $x^{(n)}(d) \ne 0$ since $x(d) \ne 0$. Thus, on $[c,d]$ we have

$$\sum_{j=1}^{n} x^{(j)}(t)\psi_j(t) = x^{(k)}(c)\psi_k(c) = x^{(n)}(d)\psi_n(d) \ne 0.$$

Hence, $x^{(k)}(c)\psi_k(c) > 0$ and $\psi_k(c) > 0$ which again as in Theorem 20.1 yields $\text{Sgn } \psi_n(t) = (-1)^{n-k}$ on $(c,d)$.

Theorem 20.3   Suppose that $f(t,\underline{x})$ is continuous on $(a,b) \times R^{q+1}$ and satisfies the Lipschitz condition (9.12). Then, each $r$ point BVP (9.1), (2.4) has a unique solution provided $a_r - a_1 < h$, where $h = \min \{ h_k : 1 \le k \le [\frac{n}{2}] \}$, and $h_k$ is the smallest positive number such that there is a solution $x(t)$ of the BVP

$$(20.12) \qquad x^{(n)} = (-1)^k [L_0 x + \sum_{j=1}^{q} L_j |x^{(j)}|]$$

$$x^{(i-1)}(0) = 0, \; 1 \le i \le n-k$$

$$x^{(i-1)}(h_k) = 0, \; 1 \le i \le k$$

with $x(t) > 0$ on $(0,h_k)$ or $h_k = \infty$ if no such solution exists. This, result is best possible.

Proof.   It suffices to show that the condition (D) of Theorem 15.5 is satisfied. Assume that $x_1(t)$ and $x_2(t)$ are distinct solutions of (9.1) on $(a,b)$ and for $0 \leq j \leq q+1$ define the functions $h_j(t)$ by

$$h_0(t) = f(t,x_1(t),x_1'(t),\ldots,x_1^{(q)}(t))$$

$$h_j(t) = f(t,x_2(t),x_2'(t),\ldots,x_2^{(j-1)}(t),x_1^{(j)}(t),\ldots,x_1^{(q)}(t)), \quad 1 \leq j \leq q$$

$$h_{q+1}(t) = f(t,x_2(t),x_2'(t),\ldots,x_2^{(q)}(t)).$$

Then, define the functions $u_j(t)$, $0 \leq j \leq q$ by

$$u_j(t) = \begin{cases} \dfrac{h_j(t) - h_{j+1}(t)}{x_1^{(j)}(t) - x_2^{(j)}(t)} & , \quad x_1^{(j)}(t) \neq x_2^{(j)}(t) \\[2em] -L_j & , \quad \phantom{,} x_1^{(j)}(t) = x_2^{(j)}(t). \end{cases}$$

It follows from the continuity of the functions involved and the Lip-schitz condition (9.12) that for each $0 \leq j \leq q$, $u_j(t)$ is measurable on $(a,b)$ and $|u_j(t)| \leq L_j$. Furthermore, the difference $x(t) = x_1(t) - x_2(t)$ is a solution of the linear equation (20.1) on $(a,b)$.

Thus, if condition (D) of Theorem 15.5 is not satisfied, then the difference $x(t) = x_1(t) - x_2(t)$ is a nontrivial solution of an equation (20.1) for a suitable $u^0 \in U$ and $x(a_i) = 0$, $1 \leq i \leq n$. In this case Sherman [9] has proved that there is a subinterval $[c,d] \subset [a_1,a_n]$ and an integer $k$ with $1 \leq k \leq n-1$ such that the BVP

$$x^{(n)} = \sum_{j=0}^{q} u_j^0(t)x^{(j)}$$

$$x^{(i-1)}(c) = 0, \quad 1 \leq i \leq k$$

$$x^{(i-1)}(d) = 0, \quad 1 \leq i \leq n-k$$

has a nontrivial solution.

It follows from replacing t by $h_k$-t that

$$x^{(n)} = (-1)^k [L_0 x + \sum_{j=1}^{q} L_j |x^{(j)}|]$$

$$x^{(i-1)}(0) = 0, \quad 1 \leq i \leq n-k$$

$$x^{(i-1)}(h_k) = 0, \quad 1 \leq i \leq k$$

and $x(t) > 0$ on $(0, h_k)$, has a solution if and only if

(20.13)
$$x^{(n)} = (-1)^{n-k} [L_0 x + \sum_{j=1}^{q} L_j |x^{(j)}|]$$

$$x^{(i-1)}(0) = 0, \quad 1 \leq i \leq k$$

$$x^{(i-1)}(h_k) = 0, \quad 1 \leq i \leq n-k$$

and $x(t) > 0$ on $(0, h_k)$, has a solution. Thus, if $h = \min\{h_k : 1 \leq k \leq [\frac{n}{2}]\}$ and $(c,d) \subset (a,b)$ with $d - c < h$, then from Theorem 20.1 for any $u \in U$ the corresponding equation (20.1) is $(k, n-k)$ disconjugate on $(c,d)$ for any $k$ with $1 \leq k \leq n-1$. This contradiction completes the proof.

**Theorem 20.4**  Suppose that $f(t, \underline{x})$ is continuous on $(a,b) \times R^{q+1}$ and satisfies the Lipschitz condition (9.12). Then, each right $(m_1, \ldots, m_r)$ focal point BVP (9.1), (15.6) has a unique solution provided $a_r - a_1 < \delta$, where $\delta = \min\{\delta_k : 1 \leq k \leq n-1\}$, and for $1 \leq k \leq [\frac{n}{2}]$ the $\delta_k$ is the smallest positive number such that there is a solution $x(t)$ of (20.12) satisfying $x^{(i-1)}(0) = 0$, $1 \leq i \leq n-k$, $x^{(i-1)}(\delta_k) = 0$, $n-k+1 \leq i \leq n$ with $x(t) > 0$ on $(0, \delta_k)$ or $\delta_k = \infty$ if no such solution exists, whereas for $[\frac{n}{2}] + 1 \leq k \leq n-1$ the $\delta_k$ is the smallest positive number such that there is a solution $x(t)$ of (20.13) satisfying $x^{(i-1)}(0) = 0$, $n-k+1 \leq i \leq n$, $x^{(i-1)}(\delta_k) = 0$, $1 \leq i \leq n-k$ with $x(t) > 0$ on $(0, \delta_k)$ or $\delta_k = \infty$ if no such solution exists. This result is best possible.

Proof. It suffices to show that the condition $(D_2)$ of Theorem 15.9 is satisfied. As in Theorem 20.3 if $x_1(t) \not\equiv x_2(t)$, then $x(t) = x_1(t) - x_2(t)$ is a nontrivial solution of an equation from the collection (20.1) with $x^{(i-1)}(a_i) = 0$, $1 \le i \le n$. It then follows from proposition 1 of [7] that for each equation from the collection (20.1) there is a nontrivial solution $y(t)$, a $\bar{t}$ with $a_1 < \bar{t} \le a_n$, and an integer k with $1 \le k \le n$ such that $y^{(i-1)}(a_1) = 0$, $1 \le i \le k$, $y^{(i-1)}(\bar{t}) = 0$, $k+1 \le i \le n$. However, from our choice of $\delta$ and from Theorem 20.2, it follows that this is impossible if $a_n - a_1 < \delta$. This completes the proof of our theorem.

## COMMENTS AND BIBLIOGRAPHY

For the equation $x^{(4)} = f(t,x)$, Theorem 20.3 ensures that all $r(2 \le r \le 4)$ point BVPs have unique solutions provided $a_r - a_1 < h$, where $h = \min\{h_1, h_2\}$. A simple computation provides that $h_1$ is the first positive root of the equation $\tan \frac{1}{\sqrt{2}} L_0^{\frac{1}{4}} h_1 = \tanh \frac{1}{\sqrt{2}} L_0^{\frac{1}{4}} h_1$ (given by $h_1^4 L_0 = 950.8820...$), whereas $h_2$ is the first positive root of the equation $\cos L_0^{\frac{1}{4}} h_2 \cosh L_0^{\frac{1}{4}} h_2 = 1$ (given by $h_2^4 L_0 = 500.5639...$). Thus, in particular the BVP $x^{(4)} = f(t,x)$, $x(a_1) = A$, $x'(a_1) = B$, $x''(a_1) = C$, $x(a_2) = D$ has a unique solution provided $a_2 - a_1 < h$, where $h^4 L_0 = 500.5639...$. However, an application of Corollary 12.6 implies that this BVP has a unique solution provided $a_2 - a_1 < h_1$. Hence, a subclassification of r point (right $(m_1, ..., m_r)$ focal point) problems and a fresh look of Control Theory Methods would be desirable. Theorems 20.1-20.4 are adapted from the work of Jackson [2,3]. However, this technique was first used for linear problems in [5,6].

1.  Hinton, D. "Disconjugate properties of a system of differential equations", J. Diff. Equs. 2, 420-437 (1966).
2.  Jackson, L. K. "Existence and uniqueness of solutions of boundary value problems for Lipschitz equations", J. Diff. Equs. 32, 76-90 (1979).

3.  Jackson, L. K. "Boundary value problems for Lipschitz equations", Differential Equations, Eds. S. Ahmed, M. Keener and A. C. Lazer, Academic Press, New York, 1980, 31-50.

4.  Lee, E. and Markus, L. Foundations of Optimal Control Theory, Wiley, New York, 1967.

5.  Melentsova, Yu. and Mil'shtein, G. "An optimal estimate of the interval on which a multipoint boundary value problem has a solution" , Differential'nye Uravrnenija 10, 1630-1641 (1974).

6.  Melentsova, Yu. "A best possible estimate of the nonoscillation interval for a linear differential equation with coefficients bounded in $L_r$", Differencial'nye Uravrnenija 13, 1776-1786 (1977).

7.  Muldowney, J. "A necessary and sufficient condition for disfocality", Proc. Amer. Math. Soc. 74, 49-55 (1979).

8.  Peterson, A. C. "Comparison theorems and existence theorems for ordinary differential equations", J. Math. Anal. Appl. 55, 773-784 (1976).

9.  Sherman, T. "Properties of solutions of nth order linear differential equations", Pacific J. Math. 15, 1045-1060 (1965).

# 21. MATCHING METHODS

This technique is useful in proving the existence of solutions of a given BVP with the help of solutions of several other related BVPs. The main idea of this method is contained in the following : Let $x_1(t)$ be a solution of a BVP on [a,b] and $x_2(t)$ be a solution of a BVP on [b,c], where b e (a,c) and fixed, then the function $x(t)$ defined by

$$x(t) = \begin{cases} x_1(t), \ t \ e \ [a,b] \\ \\ x_2(t), \ t \ e \ [b,c] \end{cases}$$

is a solution of the given BVP on [a,c].

First, we shall apply this method for three point BVPs which are not necessarily the same as considered earlier.

**Theorem 21.1**   Suppose that

(i)   for each m e R there exist solutions to each of the four BVPs : (9.1) together with

$(21.1)_i$   $\begin{cases} x(a_1) = A_1, \ x^{(i)}(a_2) = m, \ x^{(n-1)}(a_2) = A_{n-1}, \ i = n-3 \text{ or } n-2 \\ \\ x^{(j)}(a_2) = A_{j+2}, \ 0 \le j \le n-4 \text{ if } n > 3 \end{cases}$

$(21.2)_i$   $\begin{cases} x^{(n-1)}(a_2) = A_{n-1}, \ x^{(i)}(a_2) = m, \ x(a_3) = A_n, \ i = n-3 \text{ or } n-2 \\ \\ x^{(j)}(a_2) = A_{j+2}, \ 0 \le j \le n-4 \text{ if } n > 3 \end{cases}$

(ii)   for each m e R and each $t_1$ there exists at most one solution to each of the two BVPs : (9.1) together with

$$(21.3) \quad \begin{cases} x(a_1) = A_1, \ x^{(n-2)}(t_1) = m, \ x^{(n-1)}(a_2) = A_{n-1}, \ t_1 \in (a_1, a_2] \\ \\ x^{(j)}(a_2) = A_{j+2}, \ 0 \le j \le n-4 \text{ if } n > 3 \end{cases}$$

$$(21.4) \quad \begin{cases} x^{(n-1)}(a_2) = A_{n-1}, \ x^{(n-2)}(t_1) = m, \ x(a_3) = A_n, \ t_1 \in [a_2, a_3) \\ \\ x^{(j)}(a_2) = A_{j+2}, \ 0 \le j \le n-4 \text{ if } n > 3. \end{cases}$$

Then, there exists a unique solution of the BVP : (9.1) together with

$$(21.5) \quad \begin{cases} x(a_1) = A_1, \ x^{(n-1)}(a_2) = A_{n-1}, \ x(a_3) = A_n \\ \\ x^{(j)}(a_2) = A_{j+2}, \ 0 \le j \le n-4 \text{ if } n > 3. \end{cases}$$

Proof. By taking $t_1 = a_2$ in (ii) we see that respective solutions $x_1(t,m)$ and $x_2(t,m)$ of the BVPs (9.1), $(21.1)_{n-2}$ and (9.1), $(21.2)_{n-2}$ exist and are unique. We shall first show that $x_1^{(n-3)}(a_2,m)$ is continuous and a strictly increasing function of m and its range is all of R.

Let $m_2 > m_1$ and consider $y(t) = x_1(t,m_2) - x_1(t,m_1)$. Obviously, $y^{(n-2)}(t) > 0$ for all $t \in (a_1, a_2]$, since otherwise $y^{(n-2)}(p) = 0$ for some $p \in (a_1, a_2)$ which contradicts (ii). Also, $y(a_1) = 0$, $y^{(n-2)}(t) > 0$, $t \in (a_1, a_2]$ and $y^{(j)}(a_2) = 0$, $0 \le j \le n-4$ if $n > 3$ imply that $y^{(n-3)}(a_2) > 0$. Hence, $x_1^{(n-3)}(a_2,m)$ is a strictly increasing function of m. Suppose $x_1^{(n-3)}(a_2,m)$ has a discontinuity at $m = m_1$ such that $x_1^{(n-3)}(a_2,m_1^-) = \alpha$, $x_1^{(n-3)}(a_2,m_1) = \beta$, $x_1^{(n-3)}(a_2,m_1^+) = \gamma$, then from monotonicity we have $\alpha \le \beta \le \gamma$, $\alpha < \gamma$. Let $\beta_1$ be a real number different from $\beta$ such that $\alpha < \beta_1 < \gamma$, and consider the solution $x(t)$ of the problem (9.1), $(21.1)_{n-3}$ where $x^{(n-3)}(a_2) = \beta_1$. By the hypothesis (i), $x(t)$ and all its derivatives upto the nth order exist and are well defined in $[a_1, a_2]$. In particular, $x^{(n-2)}(a_2)$ exists and has a real value, say, k. Then, $x(t)$

is identical with $x_1(t,k)$ of (9.1), (21.1)$_{n-2}$ with m = k and therefore $x_1^{(n-3)}(a_2,k) = \beta_1$ which is impossible. Thus, $x_1^{(n-3)}(a_2,m)$ is a strictly increasing continuous function of m.

To prove that $x_1^{(n-3)}(a_2,m)$ has as its range the set of all reals, let us assume that for all real m, $x_1^{(n-3)}(a_2,m)$ is bounded above, i.e., $x_1^{(n-3)}(a_2,m) \leq M < \infty$. From (i), the BVP (9.1), (21.1)$_{n-3}$ with m = M + 1 has a solution x(t) such that $x^{(n-3)}(a_2) = M + 1$. If we set $x^{(n-2)}(a_2)$ = k, we find as before that $x_1^{(n-3)}(a_2,k) = M + 1$ which contradicts our assumption on the upper bound. Similarly, $x_1^{(n-3)}(a_2,m)$ is not bounded below either.

An exact parallel treatment shows that $x_2^{(n-3)}(a_2,m)$ is a strictly decreasing continuous function of m, the range being the set of all reals. Consequently, there exists a unique $m_0$ such that $x_1^{(n-3)}(a_2,m_0)$ = $x_2^{(n-3)}(a_2,m_0)$. Thus, x(t) defined as

$$(21.6) \qquad x(t) = \begin{cases} x_1(t,m_0), \ t \ \epsilon \ [a_1,a_2] \\ \\ x_2(t,m_0), \ t \ \epsilon \ [a_2,a_3] \end{cases}$$

where $x_1^{(n-2)}(a_2,m_0) = x_2^{(n-2)}(a_2,m_0) = m_0$ is a solution of the BVP (9.1), (21.5).

To establish uniqueness, suppose y(t) is another solution distinct from x(t) in (21.6). Let the restrictions of y(t) to the subintervals $[a_1,a_2]$ and $[a_2,a_3]$ be labelled $y_1(t)$ and $y_2(t)$ respectively. Then, from hypothesis (ii), $y_1(t) = x_1(t,m^*)$ and $y_2(t) = x_2(t,m^*)$ where $m^* = x^{(n-2)}(a_2)$. If $m^* > m_0$, then the preceding proof implies that
$$y_1^{(n-3)}(a_2) = x_1^{(n-3)}(a_2,m^*) > x_1^{(n-3)}(a_2,m_0) = x_2^{(n-3)}(a_2,m_0) > x_2^{(n-3)}(a_2,m^*)$$
= $y_2^{(n-3)}(a_2)$ which is a contradiction. Thus, m* cannot be greater than $m_0$ and likewise m* cannot be less than $m_0$. Hence, m* = $m_0$, i.e., y(t) = x(t).

**Theorem 21.2**  Let $\mu, \nu \in \{0, 1, \ldots, n-2\}$. For specific values of $\mu$ and $\nu$ suppose that

     (i)   for each $m \in R$ there exist solutions to each of the four BVPs : (9.1) together with

$(21.7)_i$    $x^{(\mu)}(a_1) = A_1$, $x^{(j)}(a_2) = A_{j+2}$, $0 \le j \le n-3$, $x^{(i)}(a_2) = m$, $i = n-2$ or $n-1$

$(21.8)_i$    $x^{(j)}(a_2) = A_{j+2}$, $0 \le j \le n-3$, $x^{(i)}(a_2) = m$, $x^{(\nu)}(a_3) = A_n$, $i = n-2$ or $n-1$

     (ii)  for each $m \in R$ and $t_1$ there exists at most one solution to each of the two BVPs : (9.1) together with

$(21.9)$    $x^{(\mu)}(a_1) = A_1$, $x^{(n-1)}(t_1) = m$, $x^{(j)}(a_2) = A_{j+2}$, $0 \le j \le n-3$, $t_1 \in (a_1, a_2$

$(21.10)$    $x^{(j)}(a_2) = A_{j+2}$, $0 \le j \le n-3$, $x^{(n-1)}(t_1) = m$, $x^{(\nu)}(a_3) = A_n$, $t_1 \in [a_2, a_3)$.

Then, there exists a unique solution of the BVP : (9.1) together with

$(21.11)$    $x^{(\mu)}(a_1) = A_1$, $x^{(j)}(a_2) = A_{j+2}$, $0 \le j \le n-3$, $x^{(\nu)}(a_3) = A_n$.

**Proof.**  The proof is similar to that of Theorem 21.1.

**Theorem 21.3**  Let $q = n-1$ and suppose that there exists a function $g : [a_1, a_3] \times R^n \to R$ such that

     (i)   for each $x_{n-1}, y_{n-1} \in R$

$$f(t, y_0, \ldots, y_{n-2}, y_{n-1}) - f(t, x_0, \ldots, x_{n-2}, x_{n-1})$$

$$> g(t, y_0 - x_0, \ldots, y_{n-2} - x_{n-2}, y_{n-1} - x_{n-1})$$

when $t \in (a_1, a_2]$, $(-1)^{n-j-1}(x_j - y_j) \ge 0$, $0 \le j \le n-3$ and $y_{n-2} > x_{n-2}$, or when $t \in [a_2, a_3)$, $x_j \le y_j$, $0 \le j \le n-3$ and $y_{n-2} > x_{n-2}$

(ii)  there exists $\delta_1 > 0$ such that for each $0 < \delta < \delta_1$, the
initial value problem $x^{(n)} = g(t,\underline{x})$; $x^{(i)}(a_2) = 0$, $0 \le i \le n-1$,
$i \ne n-2$, $x^{(n-2)}(a_2) = \delta$ has a solution $x(t)$ such that
$x^{(n-2)}(t)$ does not change sign on $[a_1,a_3]$

(iii)  there exists $\delta_2 > 0$ such that for each $0 < \delta < \delta_2$, the
initial value problem $x^{(n)} = g(t,\underline{x})$; $x^{(i)}(a_2) = 0$, $0 \le i \le n-2$,
$x^{(n-1)}(a_2) = \delta(-\delta)$ has a solution $x(t)$ on $[a_2,a_3]([a_1,a_2])$
such that $x^{(n-1)}(t)$ does not change sign on $[a_2,a_3]([a_1,a_2])$

(iv)  for each $z \in R$

$$g(t,y_0,\dots,y_{n-2},z) \ge g(t,x_0,\dots,x_{n-2},z)$$

when  $t \in (a_1,a_2]$, $(-1)^{n-j-1}(x_j-y_j) \ge 0$, $0 \le j \le n-3$ and
$y_{n-2} > x_{n-2} \ge 0$, or when $t \in [a_2,a_3)$, $x_j \le y_j$, $0 \le j \le n-3$
and $y_{n-2} > x_{n-2} \ge 0$

(v)  solutions of initial value problems for (9.1) exist  and
are unique on $[x_1,x_3]$.

Then, for arbitrary but fixed $\mu,\nu \in \{0,1,\dots,n-2\}$ and $m \in R$ each of the
BVPs (9.1), $(21.7)_i$ and (9.1), $(21.8)_i$ has at most one solution.


Proof.  The uniqueness of the solution of (9.1), $(21.7)_{n-2}$ with $\mu = 0$
will be proved, and remaining problems can be treated in an analogous
manner.


Suppose for a fixed m there exist two solutions $x_1(t)$ and $x_2(t)$ of
the BVP (9.1), $(21.7)_{n-2}$. Let $y(t) = x_1(t) - x_2(t)$, then $y(a_1) = y^{(j)}(a_2)$
$= 0$, $0 \le j \le n-2$. From hypothesis (v), we may assume without loss of
generality that $y^{(n-1)}(a_2) < 0$. It follows that there exists a $t_1 \in$
$(a_1,a_2)$ such that $y^{(n-1)}(t_1) = 0$ and $y^{(n-1)}(t) < 0$ on $(t_1,a_2]$. From
this, it is easy to see that $(-1)^j y^{(n-j)}(t) > 0$ on $[t_1,a_2)$, $2 \le j \le n$.
Now, let $0 < \delta < \min\{\delta_2,-y^{(n-1)}(a_2)\}$ and let $y_\delta(t)$ satisfy the hypothesis
(iii) relative to the interval $[a_1,a_2]$, i.e.,

$$y_\delta^{(n)}(t) = g(t,\underline{y}_\delta(t)), \quad y_\delta^{(i)}(a_2) = 0, \quad 0 \le i \le n-2, \quad y_\delta^{(n-1)}(a_2) = -\delta$$

and $y_\delta^{(n-1)}(t)$ does not change sign on $[a_1,a_2]$.

Now, set $z(t) = y(t) - y_\delta(t)$, then $z^{(i)}(a_2) = 0$, $0 \le i \le n-2$ and $z^{(n-1)}(a_2) < 0$. Moreover, by hypothesis (iii), we have $z^{(n-1)}(t_1) \ge 0$ and hence there exists $t_1 \le t_2 < a_2$ such that $z^{(n-1)}(t_2) = 0$ and $z^{(n-1)}(t) < 0$ on $(t_2,a_2]$. Consequently, $(-1)^j z^{(n-j)}(t) > 0$ on $[t_2,a_2)$, $2 \le j \le n$.

The following contradiction arises; first

$$z^{(n)}(t_2) = \lim_{t \to t_2^+} \frac{z^{(n-1)}(t)}{t - t_2} \le 0$$

whereas, from hypothesis (i) and (iv)

$$z^{(n)}(t_2) = y^{(n)}(t_2) - y_\delta^{(n)}(t_2) > g(t_2,\underline{y}(t_2)) - g(t_2,\underline{y}_\delta(t_2)) \ge 0.$$

Thus, $x_1(t) \equiv x_2(t)$.

**Theorem 21.4**  Assume that hypotheses (i)-(v) of Theorem 21.3 are satisfied. Then, for arbitrary but fixed $\mu,\nu \in \{0,1,\dots,n-2\}$ the BVP (9.1), (21.11) has at most one solution.

**Proof.**  Only the case $\mu = \nu = 0$ will be considered, and remaining cases can be treated in an analogous manner. Let $x_1(t)$ and $x_2(t)$ be two solutions of the BVP (9.1), (21.11). Let $y(t) = x_1(t) - x_2(t)$, then $y(a_1) = y^{(j)}(a_2) = y(a_3) = 0$, $0 \le j \le n-3$. From Theorem 21.3, $y^{(n-2)}(a_2)$, $y^{(n-1)}(a_2) \ne 0$. Assume without loss of generality that $y^{(n-2)}(a_2) > 0$. Then, there exist points $a_1 < t_1 < a_2 < t_2 < a_3$ such that $y^{(n-2)}(t_i) = 0$; $i = 1,2$ and $y^{(n-2)}(t) > 0$ on $(t_1,t_2)$. Let $0 < \delta < \min\{\delta_1, y^{(n-2)}(a_2)\}$ and let $y_\delta(t)$ be a solution satisfying the hypothesis (ii) of Theorem

21.3, and set $z(t) = y(t) - y_\delta(t)$. Then, $z^{(j)}(a_2) = 0$, $0 \leq j \leq n-3$, $z^{(n-2)}(a_2) > 0$, $z^{(n-1)}(a_2) = y^{(n-1)}(a_2) \neq 0$, and $z^{(n-2)}(t_2) \leq 0$.

Now, we need to consider the following two cases :

(1)  Suppose $z^{(n-1)}(a_2) > 0$. In view of the fact that $z^{(n-2)}(a_2) > 0$ and $z^{(n-2)}(t_2) \leq 0$, there exists $a_2 < t_3 < t_2$ such that $z^{(n-1)}(t_3) = 0$ and $z^{(n-1)}(t) > 0$ on $[a_2, t_3)$. Then, $z^{(j)}(t) > 0$ on $(a_2, t_3]$, $0 \leq j \leq n-2$. Now

$$z^{(n)}(t_3) = \lim_{t \to t_3^-} \frac{z^{(n-1)}(t)}{t - t_3} \leq 0$$

whereas, from hypotheses (i) and (iv) of Theorem 21.3, we have

$$z^{(n)}(t_3) = y^{(n)}(t_3) - y_\delta^{(n)}(t_3) > g(t_3, \underline{y}(t_3)) - g(t_3, \underline{y}_\delta(t_3)) \geq 0$$

a contradiction.

(2)  Suppose $z^{(n-1)}(a_2) < 0$. This case also gives a contradiction by making the analogous argument on $[a_1, a_2]$.

Thus, the assumption concerning distinct solutions $x_1(t)$ and $x_2(t)$ is false and the proof is complete.

Theorem 21.5   Assume that hypotheses (i)-(v) of Theorem 21.3 are satisfied. Further, we assume that for arbitrary but fixed $\mu, \nu \in \{0, 1, \ldots, n-2\}$ and $m \in R$, there exist solutions to each of the four BVPs (9.1), (21.7)$_i$ and (9.1), (21.8)$_i$; $i = n-2, n-1$. Let these solutions be denoted by $x_1(t,m)$, $x_2(t,m)$ and $y_1(t,m)$, $y_2(t,m)$ for $i = n-2$ and $n-1$ respectively. Then, $x_1^{(n-1)}(a_2,m)$ and $y_1^{(n-2)}(a_2,m)$ ($x_2^{(n-1)}(a_2,m)$ and $y_2^{(n-2)}(a_2,m)$) are strictly increasing (decreasing) functions of $m$ with ranges all of $R$.

Proof.   The strictness of the conclusion follows from Theorem 21.3.  We shall show that $x_1^{(n-1)}(a_2,m)$ for $\mu = 0$ is increasing, and remaining cases can be treated in an analogous manner. Let $m_2 > m_1$ and let $y(t) = x_1(t,m_2) - x_1(t,m_1)$, then $y(a_1) = 0$, $y^{(j)}(a_2) = 0$, $0 \leq j \leq n-3$ and $y^{(n-2)}(a_2) > 0$. Obviously, from Theorem 21.3, $y^{(n-1)}(a_2) \neq 0$. Contrary to the conclusion, assume that $y^{(n-1)}(a_2) < 0$. Since there exists $a_1 < t_1 < a_2$ such that $y^{(n-2)}(t_1) = 0$ and $y^{(n-2)}(t) > 0$ on $(t_1,a_2]$, it follows by continuity that there exists $t_1 < t_2 < a_2$ such that $y^{(n-1)}(t_2) = 0$ and $y^{(n-1)}(t) < 0$ on $(t_2,a_2]$. We also have $(-1)^j y^{(n-j)}(t) > 0$ on $[t_2,a_2)$, $2 \leq j \leq n$.

Now, let $0 < \delta < \min\{\delta_2, -y^{(n-1)}(a_2)\}$ and let $y_\delta(t)$ be a solution of the initial value problem satisfying the hypothesis (iii) of Theorem 21.3. Set $z(t) = y(t) - y_\delta(t)$, then $z^{(i)}(a_2) = 0$, $0 \leq i \leq n-3$, $z^{(n-2)}(a_2) = y^{(n-2)}(a_2) > 0$ and $z^{(n-1)}(a_2) < 0$. Furthermore, $z^{(n-1)}(t_2) \geq 0$, thus there exists $t_2 \leq t_3 < a_2$ such that $z^{(n-1)}(t_3) = 0$ and $z^{(n-1)}(t) < 0$ on $(t_3,a_2]$. Then, $(-1)^j z^{(n-j)}(t) > 0$ on $[t_3,a_2)'$, $2 \leq j \leq n$. Next as in Theorem 21.3, we get $z^{(n)}(t_3) \leq 0$ and $z^{(n)}(t_3) > 0$ which is a contradiction. Thus, $y^{(n-1)}(a_2) > 0$ and consequently $x_1^{(n-1)}(a_2,m)$ is a strictly increasing function of $m$.

In order to show that $\{x_1^{(n-1)}(a_2,m) : m \in R\} = R$, let $k \in R$ and consider the solution $y_1(t,k)$ of the BVP (9.1), $(21.7)_{n-1}$ with $\mu = 0$. Consider also the solution $x_1(t,y_1^{(n-2)}(a_2,k))$ of (9.1), $(21.7)_{n-2}$ with $\mu = 0$. Then, $x_1(t,y_1^{(n-2)}(a_2,k))$ and $y_1(t,k)$ are solutions of the same type problem (9.1), $(21.7)_{n-2}$ with $\mu = 0$, and hence by Theorem 21.3, the two functions are identical. Therefore, $x_1^{(n-1)}(a_2,y_1^{(n-2)}(a_2,k)) = y_1^{(n-1)}(a_2,k) = k$ and the statement concerning the range of $x_1^{(n-1)}(a_2,m)$ is verified.

**Theorem 21.6**  Let the conditions of Theorem 21.5 be satisfied. Then, for arbitrary but fixed $\mu, \nu \in \{0,1,\ldots,n-2\}$ the BVP (9.1), (21.11) has a unique solution.

Proof. The existence is immediate from Theorem 21.5. In fact, there exists a unique $m_0 \in R$ such that $x_1^{(n-1)}(a_2,m_0) = x_2^{(n-1)}(a_2,m_0)$. Then,

$$x(t) = \begin{cases} x_1(t,m_0), & t \in [a_1,a_2] \\ x_2(t,m_0), & t \in [a_2,a_3] \end{cases}$$

is a solution of (9.1), (21.11) and by Theorem 21.4 it is unique. Obviously, the solutions $y_1(t,m)$, $y_2(t,m)$ can also be used to find the required solution of the BVP (9.1), (21.11).

Remark 21.1    The function $g \equiv 0$ obviously satisfies hypotheses (ii) - (iv) of Theorem 21.3, and therefore Theorems 21.3-21.6 generalize several results obtained in [5, 6, 9] as particular cases. Further, with this choice of g when $n = 3$, $\mu = \nu = 0$ and f satisfying the Lipschitz condition (9.12) on $[a_1,a_3] \times R^3$, Barr and Sherman [5] used Theorem 21.6 to show that the BVP (9.1), (21.11) has a unique solution provided

(21.12)
$$\frac{\sqrt{3}}{27} L_0 h_i^3 + \frac{1}{3} L_1 h_i^2 + L_2 h_i < 1$$

where $i = 1,2$ and $h_1 = a_2 - a_1$, $h_2 = a_3 - a_2$.

Theorem 21.7    Let $n = 3$, $q = 2$, $\mu = \nu = 0$ and the function f satisfy Lipschitz condition (9.12) on $[a_1,a_3] \times R^3$. Then, the BVP (9.1), (21.11) has a unique solution provided

(21.13)
$$\frac{9}{160} L_0 h_i^3 + \frac{1}{6} L_1 h_i^2 + \frac{1}{2} L_2 h_i < 1$$

where $i = 1,2$ and $h_1 = a_2 - a_1$, $h_2 = a_3 - a_2$.

Proof. The proof is an application of Theorem 21.2 and the weight function technique discussed in Section 12.

Thus, in Theorem 21.7 not only the monotonicity condition is dispensed with but also the inequality (21.12) is considerably improved.

Next, we shall apply matching technique to some $r > 2$ point BVPs using Liapunov-like functions. For this, we shall assume that

$(H_1)$   $q = n-1$ and solutions of initial value problems for (9.1) exist on $[a,c]$

$(H_2)$   there exists $F(t,x_0,x_1,\ldots,x_{n-1}) : [a,c] \times R^n \to R$ such that

(i)    $f(t,y_0,y_1,\ldots,y_{n-1}) - f(t,x_0,x_1,\ldots,x_{n-1})$

$$= F(t,y_0-x_0,y_1-x_1,\ldots,y_{n-1}-x_{n-1})$$

(ii)   $x(t) \equiv 0$ is the unique solution of the initial value problem

(21.14)                           $x^{(n)} = F(t,\underline{x})$

$$x^{(i)}(\tau) = 0, \ 0 \le i \le n-1, \ \tau \ e \ [a,c].$$

Definition 21.1   Given $M \ge 0$ and $[\alpha,\beta] \subseteq [a,b]$, where $b \ e \ (a,c)$ and fixed, a Liapunov function

$$V_M(t,x_0,x_1,\ldots,x_{n-1}) : [\alpha,\beta] \times R^n \to R$$

is a function which is continuous, locally Lipschitz with respect to $(x_0,x_1,\ldots,x_{n-1})$ and satisfies

(i)    $V_M(t,x_0,x_1,\ldots,x_{n-1}) = 0$, if $x_{n-2} = M$

(ii)   $V_M(t,x_0,x_1,\ldots,x_{n-1}) \ge 0$, if $x_{n-2} > M$.

Lemma 21.8   Suppose that $x(t)$ is a solution of (21.14) and that for some $M \ge 0$ and $[\alpha,\beta] \subseteq [a,b]$, $V_M(t,x_0,x_1,\ldots,x_{n-1})$ is a Liapunov function. Then, $V_M(t,x(t),x'(t),\ldots,x^{(n-1)}(t))$ is nondecreasing (nonincreasing) if and only if $V_M^F(t,x(t),x'(t),\ldots,x^{(n-1)}(t)) \geq 0$ $(V_M^F(t,x(t),x'(t),\ldots,$ $x^{(n-1)}(t)) \leq 0)$, where

$$V_M^F(t,x(t),x'(t),\ldots,x^{(n-1)}(t))$$

$$= \lim_{h\to 0^+} \inf \frac{1}{h}[V_M(t+h,x(t+h),x'(t+h),\ldots,x^{(n-1)}(t+h))$$

$$- V_M(t,x(t),x'(t),\ldots,x^{(n-1)}(t))].$$

Proof. The proof for the case $n = 1$ is given in Yoshizawa [12, p. 4]. The extension to the general case is trivial.

Theorem 21.9    Suppose that there exists a Liapunov function $V_0(t,x_0, x_1,\ldots,x_{n-1})$ on $[a,b]$ such that $V_0^F(t,x(t),x'(t),\ldots,x^{(n-1)}(t)) \geq 0$ for all solutions $x(t)$ of (21.14). Then, for each $m \in R$ the BVP : (9.1) together with

(21.15)        $x^{(j)}(a_i) = A_{j+1,i}$ ; $1 \leq i \leq r-1$, $0 \leq j \leq k_i$,

$$\sum_{i=0}^{r-1} k_i + (r-1) = n-1, \quad x^{(n-2)}(a_{r-1}) = m$$

where $a \leq a_1 < \ldots < a_{r-1} = b$, has at most one solution.

Proof. Let $x_1(t)$ and $x_2(t)$ be two solutions of the BVP (9.1), (21.15). Let $y(t) = x_1(t) - x_2(t)$, then $y(t)$ is a nontrivial solution of the differential equation (21.14), (as is also $-y(t)$), and satisfies $y^{(j)}(a_i) = 0$; $1 \leq i \leq r-1$, $0 \leq j \leq k_i$, $y^{(n-2)}(a_{r-1}) = 0$. It follows that there exist points $a_1 < t_1 < t_2 < t_3 \leq a_{r-1}$ such that $y^{(n-2)}(t_1)$ $= y^{(n-2)}(t_3) = y^{(n-1)}(t_2) = 0$ and $y^{(n-2)}(t)$ or $-y^{(n-2)}(t)$ has a positive local maximum at $t = t_2$. Assume without loss of generality that $y^{(n-2)}(t)$ has a positive local maximum at $t = t_2$. Now from our hypotheses $V_0(t_1,y(t_1),\ldots,y^{(n-1)}(t_1)) = V_0(t_3,y(t_3),\ldots,y^{(n-1)}(t_3)) = 0$

and $V_0(t_2, y(t_2), \ldots, y^{(n-1)}(t_2)) > 0$. However, since $V_0^F(t, y(t), \ldots, y^{(n-1)}(t)) \geq 0$, $V_0(t, y(t), \ldots, y^{(n-1)}(t))$ is nondecreasing, consequently $V_0(t_3, y(t_3), \ldots, y^{(n-1)}(t_3)) > 0$ which is a contradiction. Thus, the uniqueness statement of the theorem is true.

**Theorem 21.10** Assume that for each $m \in R$ there exists a solution $x_1(t, m)$ of the BVP : (9.1) together with

$$(21.16) \qquad x^{(j)}(a_i) = A_{j+1, i} ; \ 1 \leq i \leq r-1, \ 0 \leq j \leq k_i,$$

$$\sum_{i=0}^{r-1} k_i + (r-1) = n-1, \ x^{(n-1)}(a_{r-1}) = m$$

where $a \leq a_1 < \ldots < a_{r-1} = b$. If for each $M \geq 0$ there exists a Liapunov function $V_M(t, x_0, x_1, \ldots, x_{n-1})$ on $[a, b]$ such that $V_M^F(t, x(t), x'(t), \ldots, x^{(n-1)}(t)) \geq 0$ for all solutions $x(t)$ of (21.14), then $x_1^{(n-2)}(a_{r-1}, m)$ is a strictly increasing function of $m$.

**Proof.** Let $m_1 < m_2$ and assume $x_1^{(n-2)}(a_{r-1}, m_2) \leq x_1^{(n-2)}(a_{r-1}, m_1)$. Then, consider the nontrivial solution $y(t) = x_1(t, m_1) - x_1(t, m_2)$ of (21.14). If follows that $y^{(n-2)}(a_{r-1}) \geq 0$. Since $y^{(j)}(a_i) = 0$; $0 \leq i \leq r-1$, $0 \leq j \leq k_i$, it follows by successive application of Rolle's theorem that there exists $t_1 \in (a_1, a_{r-1})$ such that $y^{(n-2)}(t_1) = 0$. Yet, $y^{(n-2)}(t_1) = 0$, the assumption that $y^{(n-2)}(a_{r-1}) \geq 0$ and the fact that $y^{(n-1)}(a_{r-1}) < 0$ imply that there exists $t_2 \in (t_1, a_{r-1})$ such that $y^{(n-1)}(t_2) = 0$ and $y^{(n-2)}(t)$ has a positive local maximum at $t = t_2$. Now, let $t_1 < \alpha < t_2 < \beta < a_{r-1}$ be such that $y^{(n-2)}(\alpha) = y^{(n-2)}(\beta) = M$ and $y^{(n-2)}(t_2) > M$. Since there exists a Liapunov function $V_M$, we have $V_M(\alpha, y(\alpha), \ldots, y^{(n-1)}(\alpha)) = V_M(\beta, y(\beta), \ldots, y^{(n-1)}(\beta)) = 0$ and $V_M(t_2, y(t_2), \ldots, y^{(n-1)}(t_2)) > 0$. However, since $V_M(t, y(t), \ldots, y^{(n-1)}(t))$ is nondecreasing on $[a, b]$, it follows that $V_M(\beta, y(\beta),\ldots,y^{(n-1)}(\beta)) > 0$ which is a contradiction. Thus, $y^{(n-2)}(a_{r-1}) < 0$ and this completes the proof.

Theorem 21.11    Suppose that there exists a Liapunov function $W_0(t,x_0, x_1,\ldots,x_{n-1})$ on $[b,c]$ such that $W_0^F(t,x(t),x'(t),\ldots,x^{(n-1)}(t)) \geq 0$ for all solutions $x(t)$ of (21.14). Then, for each $m \in R$ the BVP : (9.1) together with

$$(21.17) \quad x^{(j)}(a_{r-1}) = \begin{cases} A_{j+1,r-1}, & 0 \leq j \leq k_{r-1} \\ B_j, & k_{r-1}+1 \leq j \leq n-3 \end{cases}, \; x^{(n-2)}(a_{r-1})=m, \; x(a_r)=A_{1,r}$$

where $b = a_{r-1} < a_r \leq c$, has at most one solution.

Proof.    The proof is similar to that of Theorem 21.9.

Theorem 21.12    Let $x_1(t,m)$ be as in Theorem 21.10 and assume that for each $m \in R$ there exists a solution $x_2(t,m)$ of the BVP : (9.1) together with

$$(21.18) \qquad x^{(j)}(a_{r-1}) = x_1^{(j)}(a_{r-1},m), \; 0 \leq j \leq n-3,$$

$$x^{(n-1)}(a_{r-1}) = m, \; x(a_r) = A_{1,r}$$

where $b = a_{r-1} < a_r \leq c$. If for each $M \geq 0$ there exists a Liapunov function $W_M(t,x_0,x_1,\ldots,x_{n-1})$ on $[b,c]$ such that $W_M^F(t,x(t),x'(t),\ldots, x^{(n-1)}(t)) \geq 0$ for all solutions $x(t)$ of (21.14), then $x_2^{(n-2)}(a_{r-1},m)$ is a strictly decreasing function of $m$.

Proof.    The proof is similar to that of Theorem 21.10.

Theorem 21.13    Assume that for each $m \in R$ there exist  solutions of the differential equation (9.1) satisfying boundary conditions (21.15) and (21.16) on $[a,b]$, and satisfying (21.17) and (21.18) on $[b,c]$. Assume moreover that the BVP for (21.14) on [a,b] satisfying

$$x^{(j)}(a_i) = 0; \ 1 \le i \le r-1, \ 0 \le j \le k_i, \ \sum_{i=0}^{r-1} k_i + (r-1) = n-1, \ x^{(n-1)}(a_{r-1}) = 0$$

and that the BVP for (21.14) on [b,c] satisfying

$$x^{(j)}(a_{r-1}) = 0; \ j = 0,1,\ldots,n-3, n-1 \ , \ x(a_r) = 0$$

have only the trivial solution. If for each $M \ge 0$ there exist Liapunov functions $V_M(t,x_0,x_1,\ldots,x_{n-1})$ and $W_M(t,x_0,x_1,\ldots,x_{n-1})$ on [a,b] and [b,c] respectively, such that $V_M^F(t,x(t),x'(t),\ldots,x^{(n-1)}(t)) \ge 0$ and $W_M^F(t,x(t),x'(t),\ldots,x^{(n-1)}(t)) \ge 0$ for all solutions $x(t)$ of (21.14), then the BVP : (9.1) together with

(21.19)      $$x^{(j)}(a_i) = A_{j+1,i} ; \ 1 \le i \le r-1, \ 0 \le j \le k_i,$$

$$\sum_{i=0}^{r-1} k_i + (r-1) = n-1, \ x(a_r) = A_{1,r}$$

has a solution on [a,c], where $a \le a_1 < \ldots < a_{r-1} = b < a_r \le c$.

Proof.     If $x_1(t,m)$ is a solution of the BVP (9.1), (21.16) then by Theorem 21.10, $x_1^{(n-2)}(a_{r-1},m)$ is a strictly increasing function of m. We contend, furthermore, that $x_1^{(n-2)}(a_{r-1},m)$ is a continuous function of m with range all of R. To see this, it suffices to show the later, i.e., $\{x_1^{(n-2)}(a_{r-1},m) : m \in R\} = R$.

Thus, let $k \in R$ and $x(t)$ be the solution of (9.1), (21.15) with $m = k$. Consider now the solution $y(t) = x(t) - x_1(t,x^{(n-1)}(a_{r-1}))$ of (21.14), since

$$y^{(j)}(a_i) = 0; \ 1 \le i \le r-1, \ 0 \le j \le k_i$$

and

$$y^{(n-1)}(a_{r-1}) = x^{(n-1)}(a_{r-1}) - x^{(n-1)}(a_{r-1}) = 0$$

by the hypotheses of the theorem, $y(t) \equiv 0$ and hence $x(t) = x_1(t, x^{(n-1)}(a_{r-1}))$. Consequently, $x_1^{(n-2)}(a_{r-1}, x^{(n-1)}(a_{r-1}))$ $= x^{(n-2)}(a_{r-1}) = k$, and it follows that $k \in \{x_1^{(n-2)}(a_{r-1}, m) : m \in R\}$. Thus, $x_1^{(n-2)}(a_{r-1}, m)$ is a strictly increasing, continuous function of m with range all of R.

Similarly, if as in Theorem 21.12, $x_2(t, m)$ is a solution of the BVP (9.1), (21.18) then it will follow that $x_2^{(n-2)}(a_{r-1}, m)$ is a strictly decreasing, continuous function of m with range all of R. Thus, there is a unique $m_0 \in R$ such that $x_1^{(n-2)}(a_{r-1}, m_0) = x_2^{(n-2)}(a_{r-1}, m_0)$. Then, the function

$$x(t) = \begin{cases} x_1(t, m_0), & t \in [a,b] \\ x_2(t, m_0), & t \in [b,c] \end{cases}$$

is a solution of the BVP (9.1), (21.19).

## COMMENTS AND BIBLIOGRAPHY

Matching methods were first used in [3] where they dealt with solutions of the BVP (7.1), (7.2) by matching solutions of initial value problems. In 1973, Barr and Sherman [5] adapted the arguments of [3] to match solutions of two point BVPs for third order equations to obtain solutions of three point problems, which were later improved in [1,2,6, 11] and generalized for the systems in [10]. Moorti and Garner [9] extended the results of [5] for nth order differential equations. Theorems 21.1 and 21.2 are adapted from [9]. Theorems 21.3 - 21.6 which are extensions of the results proved in [11] are due to Henderson [7]. Theorem 21.7 is taken from [1]. Barr and Miletta [4] used Liapunov-like functions to match solutions of (n-1) point linear BVPs with solutions of two point linear BVPs to obtain solutions of n point linear

BVPs. Theorems 21.9 - 21.13 extend  the results of [4] to nonlinear
problems are adapted from [8].

1.      Agarwal, R. P. and Krishnamoorthy, P. R. "Existence and uniqueness
        of solutions of boundary value problems for third order differen-
        tial equations", Proc. Indian Acad. Sci. 88 A , 105-113 (1979).
2.      Agarwal, R. P. "On boundary value problems for y''' = f(x,y,y',y'')",
        Bull. Inst. Math. Acad. Sinica 12, 153-157 (1984).
3.      Bailey, P. B., Shampine, L. F. and Waltman, P. E. Nonlinear Two
        Point Boundary Value Problems, Academic Press, New York, 1968.
4.      Barr, D. and Miletta, P. "An existence and uniqueness criterion
        for solutions of boundary value problems", J. Diff. Equs. 16,
        460-471 (1974).
5.      Barr, D. and Sherman, T. "Existence and uniqueness of solutions
        of three-point boundary value problems", J. Diff. Equs. 13, 197-
        212 (1973).
6.      Das, K. M. and Lalli, B. S. "Boundary value problems for
        y''' = f(x,y,y',y'')", J. Math. Anal. Appl. 81, 300-307 (1981).
7.      Henderson, J. "Three-point boundary value problems for ordinary
        differential equations by matching solutions", Nonlinear Analysis:
        Theory, Methods and Appl. 7, 411-417 (1983).
8.      Henderson, J. "Multipoint boundary value problems for ordinary
        differential equations by matching solutions", Pre-Print.
9.      Moorti, V. R. G. and Garner, J. B. "Existence-uniqueness theorems
        for three-point boundary value problems for nth-order nonlinear
        differential equations", J. Diff. Equs. 29, 205-213 (1978).
10.     Murthy, K. N. "Three point boundary value problems, existence
        and uniqueness", J. Math. Phyl. Sci. 11, 265-272 (1977).
11.     Rao, D. R. K. S., Murthy, K. N. and Rao, A. S. "On three-point
        boundary value problems associated with third order differential
        equations", Nonlinear Analysis : Theory, Methods and Appl. 5,
        669-673 (1981).
12.     Yoshizawa, T. Stability Theory by Liapunov's Second Method, Publ.
        Math. Soc. Japan 9, Tokyo, 1966.

# 22. MAXIMAL SOLUTIONS

In this section, we shall consider the question of existence of a maximal solution for the initial value problem : (9.1) together with

(22.1) $$x^{(i)}(t_0) = x_i \; ; \; i = 0,1,\ldots,n-1$$

where $t_0 \in (a,b)$ and f is assumed to be continuous on $(a,b) \times R^{q+1}$.

For the first order initial value problem

(22.2) $$x' = f(t,x)$$

(22.3) $$x(t_0) = x_0$$

where $t_0 \in (a,b)$ and f is continuous on $[a,b) \times R$, Montel [6], Peano [7] and Perron [8] have given proofs of the existence of a maximal solution $x_0(t)$ of (22.2), (22.3). By a maximal solution of (22.2), (22.3) we mean a solution $x_0(t)$ of (22.2), (22.3) on its maximal interval of existence such that if $x(t)$ is any solution, then $x(t) \leq x_0(t)$ holds on the common interval of existence.

Kamke [4] asked the same question for the initial value problem for the system

(22.4) $$u' = F(t,u)$$

(22.5) $$u(t_0) = u^0$$

where $t_0 \in (a,b)$ and F is continuous on $(a,b) \times R^m$. By a maximal solution in this case, we mean a solution $u^0(t)$ of (22.4), (22.5) on its maximal interval of existence such that if $u(t)$ is any solution, then $u_i^0(t) \geq u_i(t)$; $i = 1,\ldots,m$ holds on the common interval of existence. For the case $m = 2$, he gave a counterexample showing the non-existence of a maximal solution for (22.4), (22.5). He then showed that if, each $F_i$ is nondecreasing in $u_j$, $j \neq i$ then the problem (22.4), (22.5) has a maximal solution.

There are obviously other ways of generalizing the definition of a maximal solution, for example, in [2] the concept of a minimal solution of (22.2), (22.3) was introduced, and then sufficient conditions on the function f were imposed to ensure the existence of such solution. Lakshmikantham and Leela [5] have done the same for the problem (22.4), (22.5).

By a maximal solution of (9.1), (22.1) we mean a solution $x_0(t)$ on its right maximal interval of existence such that if $x(t)$ is any solution, then $x(t) \leq x_0(t)$ holds on the common interval of existence. For $n = 2$, $q = 1$ Walter [10] has shown by an example that the problem (9.1), (22.1) need not have a maximal solution. However, in his example a local maximal solution does exist. We say that $x_0(t)$ is a local maximal solution of (9.1), (22.1) in case there exists a $\varepsilon > 0$ such that $x_0(t)$ is a solution on $[t_0, t_0 + \varepsilon]$ and if $x(t)$ is any other solution on $[t_0, \omega^+)$, then $x(t) \leq x_0(t)$ on $[t_0, \min(\omega^+, t_0 + \varepsilon))$. For $n = 2$, $q = 1$ Bebernes and Ingram [1] have shown by means of a counterexample that the problem (9.1), (22.1) in general need not have a maximal or even a local maximal solution. However, the problem (9.1), (22.1) does have a maximal solution if each (n-1,0) BVP (9.1), (2.7) has at most one solution on (a,b). For this, we need

**Definition 22.1**  Let $t_0 \in (a,b)$ and $(x_0, x_1, \ldots, x_{n-1}) \in R^n$. The funnel of solutions $C(t_0)$ of the initial value problem (9.1), (22.1) is defined to be the set of all points $(t, \alpha_0, \alpha_1, \ldots, \alpha_{n-1})$ such that $t \geq t_0$ and there exists a solution $x(t)$ of (9.1), (22.1) such that $x^{(i)}(t) = \alpha_i$; $i = 0, 1, \ldots, n-1$. We also define the function $y_0(t)$ as

$$y_0(t) = \sup \{ \alpha_0 : (t, \alpha_0, \alpha_1, \ldots, \alpha_{n-1}) \in C(t_0),$$

$$\text{for some } \alpha_1, \alpha_2, \ldots, \alpha_{n-1} \}.$$

Lemma 22.1   The following hold

(i)    the maximal solution $x_0(t)$ of (9.1), (22.1), if it exists
       is unique

(ii)   if the maximal solution $x_0(t)$ exists on $[t_0,\omega^+)$, its maximal
       interval of existence, then $x_0(t) \equiv y_0(t)$ on $[t_0,\omega^+)$

(iii)  if the maximal solution $x_0(t)$ of (9.1), (22.1) exists on
       $[t_0,\omega^+)$, then the solution trajectory $C_0 = \{(t,\alpha_0,\alpha_1,\ldots,\alpha_{n-1}) :$
       $t \in [t_0,\omega^+)$, $x_0^{(i)}(t) = \alpha_i$ ; $i = 0,1,\ldots,n-1\}$ lies on the
       boundary of $C(t_0)$

(iv)   if $y_0(t)$ is a solution of (9.1), (22.1)  then $y_0(t)$ is the
       maximal solution

(v)    local maximal solutions are unique.


Proof.   (i) Let, if possible, $x_1(t)$ and $x_2(t)$ be two maximal solutions
with right maximal intervals of existence $[t_0,\omega_1)$ and $[t_0,\omega_2)$ respectively.  Then by the definition of maximal solution of (9.1), (22.1) we
must have $x_2(t) \leq x_1(t)$ and $x_1(t) \leq x_2(t)$ on the common interval of
existence.  Let us assume for definiteness $\omega_1 < \omega_2$.  Then, we have

$x_1(t) \equiv x_2(t)$ on $[t_0,\omega_1)$ and this implies that $\sum_{i=0}^{n-1} |x_2^{(i)}(t)| = \sum_{i=0}^{n-1} |x_1^{(i)}(t)|$

$\to \infty$ as $t \to \omega_1^-$.  This contradicts the assumption that $x_2(t)$ exists on
$[t_0,\omega_2)$.  Hence, $\omega_1 = \omega_2$ and $x_1(t) = x_2(t)$ on $[t_0,\omega_1)$.

(ii)  By the definition of $y_0(t)$, we have $x_0(t) \leq y_0(t)$ on $[t_0,\omega^+)$. We
will now show that $y_0(t) \leq x_0(t)$ on $[t_0,\omega^+)$.  Let $t_0 < t < \omega^+$ be arbitrary.  The existence of $x_0(t)$ guarantees that $y_0(t)$ is defined.  By the
definition of $x_0(t)$, we have that, if $x(t)$ is any other solution of
(9.1), (22.1) then $x(t) \leq x_0(t)$.  Thus by the definition of $y_0(t)$, we
must have $y_0(t) \leq x_0(t)$ and consequently $x_0(t) \equiv y_0(t)$ on $[t_0,\omega^+)$.

(iii)  It follows from (ii) and the definition of $y_0(t)$.

(iv)   If $x(t)$ is any solution of (9.1), (22.1) then by the definition
of $y_0(t)$, $x(t) \leq y_0(t)$ for any t in the common interval of existence of
$x(t)$ and $y_0(t)$.  Hence, if $y_0(t)$ is a solution, it is the maximal solution.

(v)  Let, if possible, $x_1(t)$, $x_2(t)$ be two local maximal solutions. Then, there exists a $\varepsilon_i > 0$ such that $x_i(t)$ is a solution of (9.1), (22.1) on $[t_0, t_0 + \varepsilon_i)$ and if $x(t)$ is any other solution on $[t_0, \omega^+)$, then $x(t) \leq x_i(t)$ on $[t_0, \min(t_0 + \varepsilon_i, \omega^+))$, for each i = 1,2.  Let $\varepsilon = \min(\varepsilon_1, \varepsilon_2)$. Then, $x_2(t) \leq x_1(t)$ on $[t_0, t_0 + \varepsilon)$, since $x_1(t)$ is a local maximal solution and $x_1(t) \leq x_2(t)$ on $[t_0, t_0 + \varepsilon)$, since $x_2(t)$ is a local maximal solution.  Hence, $x_1(t) \equiv x_2(t)$ on $[t_0, t_0 + \varepsilon)$.

**Theorem 22.2**  Assume solutions of (n-1,0) BVPs (9.1), (2.7) are unique when they exist.  Then, there exists a $\omega^+ > t_0$ such that $y_0(t)$ is a solution of (9.1), (22.1) on $[t_0, \omega^+)$ and hence is the maximal solution.

**Proof.**  Let $E_1, E_2, \ldots$ be open subsets of $E = (a,b) \times R^{q+1}$ such that

$$E = \bigcup_{j=1}^{\infty} E_j, \ \overline{E}_j \text{ is compact and } \overline{E}_j \subset E_{j+1} \text{ for each j.}$$  By the proof of Peano's existence theorem [3,p.10] and by corollary [3,p.11], there exists $\delta(j)$ such that if $(t_0, x_0, x_1, \ldots, x_{n-1}) \in \overline{E}_j$, then all solutions $x(t)$ of (9.1), (22.1) exist on $[t_0, t_0 + \delta(j)]$ and satisfy $|x^{(i)}(t)| \leq M_i(j)$, on $[t_0, t_0 + \delta(j)]$ for i = 0,1,...,n-1.

Let j(1) be so large that $(t_0, x_0, x_1, \ldots, x_{n-1}) \in \overline{E}_{j(1)}$ and $s = y_0(t_0 + \delta(j(1)))$.  By the definition of $y_0(t)$, there exists a sequence $\{x_m(t)\}$ of solutions of (9.1), (22.1) satisfying $x_m(t_0 + \delta(j(1))) \leq x_{m+1}(t_0 + \delta(j(1))) \leq s$ and $x_m(t_0 + \delta(j(1))) \to s$ as $m \to \infty$.  We also have, by the uniqueness of solutions of (n-1,0) BVPs (9.1), (2.7) that $x_m(t) \leq x_{m+1}(t)$ on $[t_0, t_0 + \delta(j(1))]$.  Since the sequences $\{x_m^{(i)}(t)\}$ are uniformly bounded and $x_m(t)$ satisfies (9.1) for all m, it follows that the sequences $\{x_m^{(i)}(t)\}$ are equicontinuous on $[t_0, t_0 + \delta(j(1))]$ for each i = 0,1,...,n-1.  Then, by using Ascoli's theorem we obtain a subsequence $\{x_{m(j)}(t)\}$ which we again call $\{x_m(t)\}$ with the property that $x_m^{(i)}(t) \to x_0^{(i)}(t)$, uniformly on $[t_0, t_0 + \delta(j(1))]$ for each i = 0,1,...,n-1 where $x_0(t)$ is a solution of (9.1), (22.1).  We will now show that $x_0(t) = y_0(t)$ on $[t_0, t_0 + \delta(j(1))]$.  By the definition of $y_0(t)$, we must have $x_0(t) \leq y_0(t)$ on $[t_0, t_0 + \delta(j(1))]$.  Suppose there exists $t_1 \in (t_0, t_0 + \delta(j(1)))$

such that $x_0(t_1) < y_0(t_1)$. Then by the definition of $y_0(t)$, there exists a solution $z(t)$ of (9.1), (22.1) such that $x_0(t_1) < z(t_1) < y_0(t_1)$. By the conditions (22.1) and the uniform convergence of $\{x_m(t)\}$, we have $x_0^{(i)}(t_0) = z^{(i)}(t_0)$; $i = 0,1,\ldots,n-2$, $x_0(t_0+\delta(j(1))) = y_0(t_0+\delta(j(1))) \geq z(t_0+\delta(j(1)))$. Thus, $x_0(t) = z(t)$ for some $t \in (t_1,t_0+\delta(j(1)))$ which contradicts the uniqueness of solutions of $(n-1,0)$ BVPs (9.1), (2.7) on $[t_0,t_0+\delta(j(1))]$. Now by Lemma 22.1, it follows that $y_0(t)$ is the maximal solution of (9.1), (22.1) on $[t_0,t_0+\delta(j(1))]$. If $(t_0+\delta(j(1))$, $y_0(t_0+\delta(j(1))),\ldots, y_0^{(n-1)}(t_0+\delta(j(1)))) \in \overline{E}_{j(1)}$, then $y_0(t)$ can be extended over another interval $[t_0+\delta(j(1)), t_0+2\delta(j(1))]$ of length $\delta(j(1))$. Continuing this argument, we see that there exists an integer $k(1) \geq 1$ such that the maximal solution $y_0(t)$ exists on $[t_0,t_1=t_0+k(1)\delta(j(1))]$ and $(t_1,y_0(t_1),\ldots,y_0^{(n-1)}(t_1)) \notin \overline{E}_{j(1)}$. Let $j(2)$ be so large that

$$(t_1,y_0(t_1),\ldots,y_0^{(n-1)}(t_1)) \in \overline{E}_{j(2)}.$$

Next, consider the initial value problem : (9.1) together with

(22.6)
$$x^{(i)}(t_1) = y_0^{(i)}(t_1); \quad i = 0,1,\ldots,n-1.$$

There exists a $\delta(j(2))$ such that all solutions of (9.1), (22.6) exist on $[t_1,t_1+\delta(j(2))]$ and are such that their derivatives upto order $(n-1)$ are uniformly bounded and equicontinuous by virtue of (9.1) on $[t_1,t_1+\delta(j(2))]$. Let $y_1(t) = \sup\{x(t) : x(t)$ is a solution of (9.1),(22.6) on $[t_1,t_1+\delta(j(2))]\}$. As before, we can show that $y_1(t)$ is a solution of (9.1), (22.6) on $[t_1,t_1+\delta(j(2))]$. Now, we will show that $y_0(t) \equiv y_1(t)$ on $[t_1,t_1+\delta(j(2))]$. If not, first we suppose there exists a $t_3 \in [t_1, t_1+\delta(j(2))]$ such that $y_0(t_3) > y_1(t_3)$. Then, there exists a solution $x(t)$ of (9.1), (22.1) such that $y_1(t_3) < x(t_3) < y_0(t_3)$ and by the definition of $y_0(t)$, we must have $x(t) \leq y_0(t)$ on $[t_0,t_3]$. If $x(t) \equiv y_0(t)$ on $[t_0,t_1]$, then $x^{(i)}(t_1) = y_0^{(i)}(t_1)$; $i = 0,1,\ldots,n-1$ which implies that $x(t)$ is also a solution of (9.1), (22.6). Then, the inequality $x(t_3) > y_1(t_3)$ yields a contradiction to the definition of $y_1(t)$. If $x(t) \not\equiv y_0(t)$ on $[t_0,t_1]$, then there exists a $t' \in [t_0,t_1]$ such that $x(t') < y_0(t')$. We define

$$z(t) = \begin{cases} y_0(t), & t \in [t_0, t_1] \\ \\ y_1(t), & t \in [t_1, t_1 + \delta(j(2))] \end{cases}$$

then, $z(t)$ is a solution of (9.1) satisfying

$$z^{(i)}(t_0) = x^{(i)}(t_0) = y_0^{(i)}(t_0); \quad i = 0, 1, \ldots, n-2$$

$$z(t') > x(t') \quad \text{and} \quad z(t_3) < x(t_3).$$

However, this yields a contradiction to the uniqueness of $(n-1,0)$ BVPs (9.1), (2.7). On the other hand, if $y_0(t_3) < y_1(t_3)$, we define

$$w(t) = \begin{cases} y_0(t), & t \in [t_0, t_1] \\ \\ y_1(t), & t \in [t_1, t_1 + \delta(j(2))]. \end{cases}$$

Then, $w(t)$ is a solution of (9.1), (22.1) satisfying $w(t_3) = y_1(t_3) > y_0(t_3)$ which contradicts the definition of $y_0(t)$. Hence, $y_0(t) \equiv y_1(t)$ on $[t_1, t_1 + \delta(j(2))]$. If $(t_1 + \delta(j(2)), y_0(t_1 + \delta(j(2)))), \ldots,$ $y_0^{(n-1)}(t_1 + \delta(j(2))) \in \bar{E}_{j(2)}$, then $y_0(t)$ can be extended as a solution over $t_0 \leq t \leq t_2$, yielding that $y_0(t)$ is the maximal solution on $[t_0, t_2]$, where $t_2 = t_1 + k(2)\delta(j(2))$ and $(t_2, y_0(t_2), \ldots, y_0^{(n-1)}(t_2)) \notin \bar{E}_{j(2)}$. Thus, $y_0(t)$ can be extended as a solution over its maximal interval of existence $[t_0, \omega^+)$ where $\omega^+ \leq b$. Hence, $y_0(t)$ is the maximal solution of (9.1), (22.1) on $[t_0, \omega^+)$.

## COMMENTS AND BIBLIOGRAPHY

For $q = n-1$, if the function $f(t, x_0, x_1, \ldots, x_{n-1})$ is nondecreasing in $x_j$ for fixed values of $(t, x_0, x_1, \ldots, x_{j-1}, x_{j+1}, \ldots, x_{n-1})$ for each j = 0,1,...,n-2 then from Kamke's theorem the initial value problem (9.1), (22.1) has the maximal solution. We believe that if all (0,n-1) BVPs (9.1), (2.8) have unique solutions on (a,b), then the maximal solution of (9.1), (22.1) exists on its left maximal interval of existence. Further, if all two point BVPs for (9.1) in which data upto and including the (n-2)th derivative are assigned at one of the two points have unique solutions on (a,b), then the maximal solution of (9.1), (22.1) exists on its right and left maximal interval of existence. Lemma 22.1 as well as Theorem 22.2 is adapted from [9].

1.  Bebernes, J. W. and Ingram, S. K. "Existence and non-existence of maximal solutions", Ann. Polon. Math. 25, 125-138 (1971).
2.  Burton, L. P. and Whyburn, W. M. "Minimal solutions of ordinary differential systems", Proc. Amer. Math. Soc. 3, 794-803 (1952).
3.  Hartman, P. Ordinary Differential Equations, Wiley, New York, 1964.
4.  Kamke, E. "Zur Theorie der Systeme gewöhnlicher Differentialglei- chungen, II", Acta Math. 58, 57-85 (1932).
5.  Lakshmikantham, V. and Leela, S. "Remarks on minimax solutions", Ann. Polon. Math. 19, 301-306 (1967).
6.  Montel, P. "Sur les suites de fonctions", Ann. École Norm. 24, 233-234 (1907).
7.  Peano, G. "Sull' integrabilità delle equazione differenziali di primo ordine", Atti R. Accad. Torino 21, 677-685 (1885/1886).
8.  Perron, O. "Ein neuer Existenzbeweis für die Integrale der Differentialgleichung y' = f(x,y)", Math. Ann. 76, 471-484 (1915).
9.  Umamaheswaram, S. "Boundary value problems for n-th order ordinary differential equations", Ph.D. Thesis, Univ. of Missouri, Columbia, 1973.
10. Walter, W. "On the non-existence of maximal solutions for hyper- bolic differential equations", Ann. Polon. Math. 19, 307-311 (1967).

# 23. MAXIMUM PRINCIPLE

'Everyone knows' that if $x \in C^{(2)}[a_1,a_2]$ and $x''(t) \geq 0$ on $(a_1,a_2)$, then x satisfies the maximum principle, i.e., if x attains its maximum at an interior point of $[a_1,a_2]$, then x is identically constant on $[a_1,a_2]$. This principle, however, is not true for functions satisfying higher order inequalities. For example, let $x = -t^2$, it follows that x satisfies the inequality $x^{(4)}(t) \geq 0$ on $[-1,1]$, and yet x assumes its maximum at $t = 0$.

In [1] it is proved that if $x \in C^{(4)}[a_1,a_2]$ and

$$x^{(4)}(t) \geq 0, \quad t \in (a_1,a_2)$$

$$x'(a_1) \geq 0, \quad x'(a_2) \leq 0$$

then x attains its minimum at $a_1$ or $a_2$. An alternative proof of this principle has been given in [2]. Its generalizations are given in the following :

Theorem 23.1   Let $x \in C^{(2m)}[a_1,a_2]$ and

(23.1)
$$x^{(2m)}(t) \geq 0, \quad t \in (a_1,a_2)$$

$$(-1)^m x^{(i)}(a_1) \geq 0$$

(23.2)
$$(-1)^{m+i} x^{(i)}(a_2) \geq 0; \quad i = 1,2,\ldots,m-1$$

then in the case m even (m odd) x attains its minimum (its maximum) at either $a_1$ or $a_2$.

When the inequalities in (23.1) and (23.2) are reversed, then the preceding result is true, provided the word minimum is replaced by the word maximum and vice versa.

Proof. Only the case (23.1), (23.2) and m even will be considered. We start with the integral representation

$$(23.3) \qquad x(t) = \sum_{i=0}^{m-1} q_i(t) x^{(i)}(a_1) + \sum_{i=0}^{m-1} r_i(t) (-1)^i x^{(i)}(a_2)$$

$$+ \int_{a_1}^{a_2} g_3(t,s) x^{(2m)}(s) ds$$

where $g_3(t,s)$ is defined in (3.14), and $q_i$ is the polynomial of degree 2m-1 satisfying

$$q_i^{(i)}(a_1) = 1, \quad q_i^{(j)}(a_1) = 0; \quad j = 0,1,\ldots,m-1, \; j \neq i$$

(23.4)

$$q_i^{(j)}(a_2) = 0; \quad j = 0,1,\ldots,m-1$$

and $r_i$ is the polynomial of degree 2m-1 satisfying

$$(-1)^i r_i^{(i)}(a_2) = 1, \quad r_i^{(j)}(a_2) = 0; \quad j = 0,1,\ldots,m-1, \; j \neq i$$

(23.5)

$$r_i^{(j)}(a_1) = 0; \quad j = 0,1,\ldots,m-1.$$

From the representation (3.14) or as an application of Lemma 3.2, we have

$$(23.6) \qquad g_3(t,s) \geq 0 \quad \text{for} \quad a_1 \leq s,t \leq a_2.$$

Since 1 satisfies the same boundary conditions as $q_0 + r_0$, we get

$$(23.7) \qquad q_0(t) + r_0(t) = 1, \quad t \in [a_1, a_2].$$

Now, we shall show that

$$(23.8) \qquad q_i(t) > 0; \; t \in (a_1, a_2), \; i = 0,1,\ldots,m-1.$$

First, we shall prove (23.8) for $i = m-1$. In view of (23.4), $q_{m-1}$ cannot have any further zero besides the prescribed ones at $a_1$ and $a_2$. Similarly, $q_i^{(i+1)}$, $i \in \{0,1,\ldots,m-2\}$ can have at most i zeros in $(a_1,a_2)$. Thus, $q_0$ must be strictly decreasing in $[a_1,a_2]$ and it has no zero at all in $(a_1,a_2)$. Let $i \in \{1,\ldots,m-2\}$ and suppose $q_i$ has at least one zero in $(a_1,a_2)$. Using Rolle's theorem, step by step we come to the conclusion that $q_i'$ must have at least two zeros in $(a_1,a_2),\ldots,q^{(i)}$ as well as $q_i^{(i+1)}$ must possess i+1 zeros in $(a_1,a_2)$. The obtained contradiction shows that (23.8) is true.

As to $r_i$, $i \in \{0,1,\ldots,m-1\}$, by the transformation $a_2 - t = \tau - a_2$ the interval $[a_1,a_2]$ goes into $[a_2,2a_2-a_1]$, and the boundary conditions (23.5) change into (23.4) at $a_2$ and $2a_2 - a_1$. Thus, the inequalities (23.8) imply that

$$(23.9) \qquad r_i(t) > 0 ; \; t \in (a_1,a_2), \; i = 0,1,\ldots,m-1.$$

Using (23.1), (23.2), (23.6)-(23.9) in (23.3), we find

$$x(t) \geq \min\{x(a_1),\, x(a_2)\}.$$

This completes the proof of our theorem.

Theorem 23.2    Let $x \in C^{(n)}[a_1,a_2]$ and

$$(23.10) \qquad x^{(n)}(t) \geq 0, \; t \in (a_1,a_2)$$

$$(23.11) \qquad x^{(i)}(a_1) \leq 0; \; i = 1,2,\ldots,n-2.$$

Then, x attains its maximum either at $a_1$ or at $a_2$.

When $x \in C^{(n)}[a_1,a_2]$, satisfies (23.10) and

$$(23.12) \qquad (-1)^{n-1+i} x^{(i)}(a_2) \geq 0; \; i = 1,2,\ldots,n-2$$

then in the case n odd (n even) x attains its minimum (its maximum) at $a_1$ or $a_2$.

Further, reversing the inequalities (23.10), (23.11) or (23.10), (23.12) we get from the maximum principle the minimum principle and vice versa.

Proof.   We shall prove only for the case (23.10), (23.11).   Consider the integral representation

$$(23.13) \quad x(t) = \sum_{i=0}^{n-2} q_i(t) x^{(i)}(a_1) + r_0(t) x(a_2) + \int_{a_1}^{a_2} g_5(t,s) x^{(n)}(s) ds$$

where $g_5(t,s)$ is defined in (3.21), and

$$(23.14) \quad q_i(t) = \frac{(t-a_1)^i}{i!} [1 - (\frac{t-a_1}{a_2-a_1})^{n-i-1}]; \quad i = 0,1,\ldots,n-2$$

$$(23.15) \quad r_0(t) = (\frac{t-a_1}{a_2-a_1})^{n-1}.$$

From Lemma 3.6, we have

$$(23.16) \quad g_5(t,s) \leq 0, \quad a_1 \leq s,t \leq a_2.$$

Further, from (23.14) and (23.15) it is obvious that

$$(23.17) \quad q_0(t) + r_0(t) = 1, \quad t \in [a_1,a_2]$$

$$(23.18) \quad q_i(t) > 0; \quad t \in (a_1,a_2), \quad i = 0,1,\ldots,n-2$$

$$(23.19) \quad r_0(t) > 0, \quad t \in (a_1,a_2).$$

Using (23.11) and (23.16)-(23.19) in (23.13), we get

$$x(t) \leq \max\{x(a_1),x(a_2)\}.$$

As an application of Theorem 23.1, we shall prove the following :

**Theorem 23.3**    Suppose that $n = 2m$ and $m$ is even and that $f(t,\underline{x})$ satisfies the inequality

$$(23.20) \qquad 0 \le f(t,x_0,x_1,\ldots,x_q) \le L + \sum_{i=0}^{q} L_i |x_i|$$

for $x_0 \ge 0$ and arbitrary $t \in [a_1,a_2]$ and $x_1,\ldots,x_q$. Then, there exists a nonnegative solution of the differential equation (9.1), satisfying the boundary conditions

$$(23.21) \qquad x^{(i)}(a_1) = A_i \ge 0, \ (-1)^i x^{(i)}(a_2) = B_i \ge 0; \ i = 0,1,\ldots,m-1$$

provided $\theta < 1$, where $\theta$ is defined in (9.8).

Proof.    The function

$$g(t,x_0,x_1,\ldots,x_q) = f(t,|x_0|,x_1,\ldots,x_q)$$

satisfies the inequality

$$0 \le g(t,x_0,x_1,\ldots,x_q) \le L + \sum_{i=0}^{q} L_i |x_i|.$$

Thus, from Theorem 9.10 the BVP : $x^{(2m)} = g(t,\underline{x})$ together with (23.21) has a solution $x(t)$. Since $g$ is nonnegative, the solution $x(t)$ satisfies all the assumptions of Theorem 23.1. Thus, $x(t) \ge \min\{x(a_1),x(a_2)\} = \min\{A_1,B_1\} \ge 0$ for all $t \in (a_1,a_2)$, and this implies that $x(t)$ is a solution of (9.1), (23.21).

## COMMENTS AND BIBLIOGRAPHY

Maximum principles which are known for ordinary and partial differential equations find a key position in existence and uniqueness theory, and in the construction of the solutions [3,4]. Theorems 23.1

and 23.2 are adapted from the work of Šeda [5].

1. Chow, Shui-Nee, Dunninger, D. R. and Lasota, A. "A maximum principle for fourth order ordinary differential equations", J. Diff. Equs. 14, 101-105 (1973).
2. Kuttler, J. R. "A remark on the paper "A maximum principle for fourth order ordinary differential equations", by Chow, Dunninger, and Lasota", J. Diff. Equs. 17, 44-45 (1975).
3. Protter, M. H. and Weinberger, H. F. Maximum Principles in Differential Equations, Prentice-Hall, Englewood Cliffs, N. J. 1967.
4. Sattinger, D. H. Topics in Stability and Bifurcation Theory, Lecture notes in Mathematics 309, Springer-Verlag, New York, 1973.
5. Šeda, V. "Two remarks on boundary value problems for ordinary differential equations", J. Diff. Equs. 26, 278-290 (1977).

## 24. INFINITE INTERVAL PROBLEMS

In this section, we are concerned with the existence of solutions of BVPs on semi-infinite and infinite intervals. In Theorems 24.2 and 24.3, we shall provide sufficient conditions for the BVP : (9.1) with q = 0 and

$$(24.1) \qquad x^{(j)}(a_i) = A_{j+1,i}; \ 1 \le i \le r_1, \ 0 \le j \le \ell_i, \ \sum_{i=1}^{r_1} \ell_i + r_1 = n-1$$

where $a_1 < \ldots < a_{r_1}$ $(r_1 \ge 2)$, to have a solution on $[a_1, \infty)$ and $(-\infty, a_{r_1}]$

respectively. In Theorem 24.4 sufficient conditions are given for the existence of a solution on $(-\infty, \infty)$ of the BVP : (9.1) with q = 0 and

$$(24.2) \qquad x^{(j)}(a_i) = A_{j+1,i}; \ 1 \le i \le r_2, \ 0 \le j \le m_i, \ \sum_{i=1}^{r_2} m_i + r_2 = n-2$$

where $a_1 < \ldots < a_{r_2}$ $(r_2 \ge 2)$. For this, we need the following :

Lemma 24.1   Suppose that q = 0 and the following hold

(i)    there exist lower and upper solutions $x_0(t)$, $y_0(t)$ of (9.1) such that

$$\text{Sgn}(x_0(t) - y_0(t)) = (-1)\exp[n + \sum_{j=1}^{i} k_j + i]; \ a_i < t < a_{i+1},$$

$$1 \le i \le r-1$$

(ii)   $x_0^{(j)}(a_i) = y_0^{(j)}(a_i) = A_{j+1,i}; \ 2 \le i \le r-1, \ 0 \le j \le k_i$

$x_0^{(j)}(a_i) = y_0^{(j)}(a_i) = A_{j+1,i}; \ i = 1, r, \ 0 \le j \le k_i - 1 \ (\text{if } k_i \ge 1)$

(iii)  $(-1)^{n+k_1}(x_0^{(k_1)}(a_1) - A_{k_1+1,1}) \le 0 \le (-1)^{n+k_1}(y_0^{(k_1)}(a_1) - A_{k_1+1,1})$

(iv)   $x_0^{(k_r)}(a_r) \le A_{k_r+1,r} \le y_0^{(k_r)}(a_r)$

(v) $\quad y_0^{(n)}(t) \le f(t,x) \le x_0^{(n)}(t)$

for all $(t,x) \in [a_1, a_r] \times W$, where

$W = \{x : x_0(t) \le x \le y_0(t)\} \cup \{x : y_0(t) \le x \le x_0(t)\}$.

Then, the BVP (9.1), (2.4) has a solution $x(t)$ such that $x(t) \in W$.

**Proof.** Theorem 14.4 implies that it is sufficient to show that the condition (ii) of Theorem 14.3 is satisfied. For this, the definition of partial ordering in $C[a_1, a_r]$ implies that inequality (14.3) is equivalent to condition (i). To prove the inequality (14.4), from the explicit representation of $P_{n-1}(t)$ and condition (ii), we have

$$P_{n-1}(t) - P_{n-1,x_0}(t)$$

$$= \frac{1}{k_1!} \frac{(t-a_1)^{k_1}(t-a_2)^{k_2+1} \cdots (t-a_r)^{k_r+1}}{(a_1-a_2)^{k_2+1} \cdots (a_1-a_r)^{k_r+1}} (A_{k_1+1,1} - x_0^{(k_1)}(a_1))$$

$$+ \frac{1}{k_r!} \frac{(t-a_1)^{k_1+1} \cdots (t-a_{r-1})^{k_{r-1}+1}(t-a_r)^{k_r}}{(a_r-a_1)^{k_1+1} \cdots (a_r-a_{r-1})^{k_{r-1}+1}} (A_{k_r+1,r} - x_0^{(k_r)}(a_r)).$$

Let $t \in [a_i, a_{i+1}]$ and $i \in H_1$, then from (iii) and (iv) we have

$$Sgn(P_{n-1}(t) - P_{n-1,x_0}(t)) = (-1)^{n+k_1+1}(-1)^{odd} Sgn(A_{k_1+1,1} - x_0^{(k_1)}(a_1))$$

$$+ (-1)^{even} Sgn(A_{k_r+1,1} - x_0^{(k_r)}(a_r))$$

$$= (-1)^{even}$$

and hence $P_{n-1}(t) \ge P_{n-1,x_0}(t)$. This, together with the reverse inequality in the interval $[a_i, a_{i+1}]$ for $i \in H_2$ gives $P_{n-1} \ge P_{n-1,x_0}$. The

Proof for $P_{n-1} \leq P_{n-1,y_0}$ is similar.

Theorem 24.2  Suppose that $q = 0$ and the following hold

(i)   there exist lower and upper solutions $x_0(t)$, $y_0(t)$ of (9.1) on $[a_1,\infty)$ such that

$$\text{Sgn}(x_0(t)-y_0(t)) = (-1)\exp[n + \sum_{j=1}^{i} \ell_j+1]; \quad a_i < t < a_{i+1},$$

$$1 \leq i \leq r_1-1$$

(ii)  $x_0^{(j)}(a_i) = y_0^{(j)}(a_i) = A_{j+1,i}; \quad 2 \leq i \leq r_1, \ 0 \leq j \leq \ell_i$

$x_0^{(j)}(a_1) = y_0^{(j)}(a_1) = A_{j+1,1}, \quad 0 \leq j \leq \ell_1-1 \quad (\text{if } \ell_1 \geq 1)$

(iii) $(-1)^{n+\ell_1}(x_0^{(\ell_1)}(a_1) - A_{\ell_1+1,1}) \leq 0 \leq (-1)^{n+\ell_1}(y_0^{(\ell_1)}(a_1)-A_{\ell_1+1,1})$

(iv)  $x_0(t) < y_0(t), \ a_{r_1} < t$

(v)   $y_0^{(n)}(t) \leq f(t,x) \leq x_0^{(n)}(t)$

for all $(t,x) \in [a_1,\infty) \times W$, where

$$W = \{x : x_0(t) \leq x \leq y_0(t)\} \cup \{x : y_0(t) \leq x \leq x_0(t)\}.$$

Then, the BVP (9.1), (24.1) has a solution $x(t)$ on $[a_1,\infty)$ such that $x(t) \in W$.

Proof.  Let $\{t_i\}$ be a sequence such that $t_i = a_i$, $1 \leq i \leq r_1$ and $t_{r_1} < t_{r_1+1} < \ldots \to \infty$. For $i \geq r_1+1$, let $x_i(t)$ be a solution of (9.1) with $q = 0$, satisfying the boundary conditions (24.1) and $x_i(t_i) = \dfrac{x_0(t_i)+y_0(t_i)}{2}$. The existence of $x_i(t)$ for each $i \geq r_1+1$ on $[a_1,t_i]$ is guaranteed by Lemma 24.1, and $x_i(t) \in W$. This together with the condition (v) implies that there exists a positive number M such that

$|x_i(t)| \leq M$ and $|x_i^{(n)}(t)| \leq M$ on $[a_1,a_2]$ for all $i \geq r_1+1$. Hence, by

Lemma 14.2 there exists a subsequence of $\{x_i(t)\}$ denoted again by

$\{x_i(t)\}$ such that $\{x_i^{(j)}(t)\}$ converges uniformly on $[a_1,a_2]$ for each

$j$, $0 \leq j \leq n-1$. Now, since $\{x_i^{(j)}(a_1)\}$ converges for each $j$, $0 \leq j \leq n-1$

by Theorem 15.1 there exists a subsequence of solutions, again denoted

by $\{x_i(t)\}$ converging uniformly to some solution $x(t)$ of (9.1) with

$q = 0$ on compact subsets of $[a_1,\infty)$ and $x(t)$ satisfies the boundary con-

ditions (24.1), since all the $x_i(t)$ satisfy them. Further, $x(t) \in W$.

This completes the proof.

Theorem 24.3   Suppose that $q = 0$ and the following hold

(i)   there exist lower and upper solutions $x_0(t)$, $y_0(t)$ of (9.1)

on $(-\infty, a_{r_1}]$ such that

$$\text{Sgn}(x_0(t) - y_0(t)) = (-1)\exp[n + \sum_{j=1}^{i} \ell_j + i + 1]; \quad a_i < t < a_{i+1},$$

$$1 \leq i \leq r_1-1$$

(ii)   $x_0^{(j)}(a_i) = y_0^{(j)}(a_i) = A_{j+1,i}; \quad 1 \leq i \leq r_1-1, \ 0 \leq j \leq \ell_i$

$x_0^{(j)}(a_{r_1}) = y_0^{(j)}(a_{r_1}) = A_{j+1,r_1}, \quad 0 \leq j \leq \ell_{r_1}-1 \ (\text{if } \ell_{r_1} \geq 1)$

(iii)   $x_0^{(\ell_{r_1})}(a_{r_1}) \leq A_{\ell_{r_1}+1,r_1} \leq y_0^{(\ell_{r_1})}(a_{r_1})$

(iv)   $(-1)^n x_0(t) < (-1)^n y_0(t)$, $t < a_1$

(v)   $y_0^{(n)}(t) \leq f(t,x) \leq x_0^{(n)}(t)$

for all $(t,x) \in (-\infty, a_{r_1}] \times W$, where

$W = \{x : x_0(t) \leq x \leq y_0(t)\} \cup \{x : y_0(t) \leq x \leq x_0(t)\}$.

Then, the BVP (9.1), (24.1) has a solution $x(t)$ on $(-\infty, a_{r_1}]$ such that

$x(t) \in W$.

Proof. The proof is similar to that of Theorem 24.2.

Theorem 24.4    Suppose that $q = 0$ and the following hold

(i)    there exist lower and upper solutions $x_0(t)$, $y_0(t)$ of (9.1) on $(-\infty, \infty)$ such that

$$\text{Sgn}(x_0(t) - y_0(t)) = (-1)\exp[n + \sum_{j=1}^{i} m_j + i + 1]; \quad a_i < t < a_{i+1},$$
$$1 \le i \le r_2 - 1$$

(ii)    $x_0^{(j)}(a_i) = y_0^{(j)}(a_i) = A_{j+1,i}; \quad 1 \le i \le r_2, \quad 0 \le j \le m_i$

(iii)    $(-1)^n x_0(t) < (-1)^n y_0(t)$, $t < a_1$

(iv)    $x_0(t) < y_0(t)$, $a_{r_2} < t$

(v)    $y_0^{(n)}(t) \le f(t,x) \le x_0^{(n)}(t)$

for all $(t,x) \in (-\infty, \infty) \times W$, where

$$W = \{x : x_0(t) \le x \le y_0(t)\} \cup \{x : y_0(t) \le x \le x_0(t)\}.$$

Then, the BVP (9.1), (24.2) has a solution $x(t)$ on $(-\infty, \infty)$ such that $x(t) \in W$.

Proof.  Let $\{t_{(-i)}\}$ be a sequence of real numbers satisfying $a_1 > t_{(-1)} > t_{(-2)} > \ldots \to -\infty$ and $x_i(t)$ be a solution on $[t_{(-i)}, \infty)$ of (9.1) with $q = 0$, satisfying (24.2) and $x_i(t_{(-i)}) = \dfrac{x_0(t_{(-i)}) + y_0(t_{(-i)})}{2}$.  The existence of $x_i(t)$ for each $i \ge 1$ on $[t_{(-i)}, \infty)$ is guaranteed by Theorem 24.2, and $x_i(t) \in W$.  This together with the condition (v) implies that there exists a positive number M such that $|x_i(t)| \le M$ and $|x_i^{(n)}(t)| \le M$ on $[a_1, a_2]$ for all $i \ge 1$.  Hence, by Lemma 14.2 there exists a sub-sequence of $\{x_i(t)\}$ denoted again by $\{x_i(t)\}$ such that $\{x_i^{(j)}(t)\}$ converges for each $j$, $0 \le j \le n-1$.  Now, since $\{x_i^{(j)}(a_1)\}$ converges for each j, $0 \leq j \leq n-1$ by Theorem 15.1 there exists a subsequence of solutions, again denoted by $\{x_i(t)\}$ converging uniformly to some solution $x(t)$ of (9.1) with $q = 0$ on compact subsets of $(-\infty,\infty)$ and $x(t)$ satisfies the boundary conditions (24.2), since all the $x_i(t)$ satisfy them. Further, $x(t) \in W$. This completes the proof of our theorem.

## COMMENTS AND BIBLIOGRAPHY

Second order infinite interval problems

$$\left\{ \begin{array}{l} x'' = f(t,x) \\ x(0) = -\alpha, \ \alpha > 0 \\ x'(t) \geq 0, \ x(t) \leq 0 \ \text{on} \ [0,\infty) \end{array} \right.$$

$$\left\{ \begin{array}{l} x'' = f(t,x) \\ x(0) = \alpha, \ \alpha \ \text{real} \\ x(t) \ \text{bounded on} \ [0,\infty) \end{array} \right.$$

and

$$\left\{ \begin{array}{l} x'' = f(t,x) \\ x(t) \ \text{bounded on} \ (-\infty,\infty) \end{array} \right.$$

have been studied extensively [1-11,14]. However, for the nth order differential equations not much is known. Theorems 24.2 - 24.4 are modelled after Umamaheswaram [12]. In [13] sufficient conditions for the existence of a solution $x(t)$ on $(-\infty,\infty)$ of the BVP : (9.1) with $q = 0$, (24.2) which in addition satisfies $x(a_0) = A_0$, where $a_0 \neq a_i$, $1 \leq i \leq r_2$ or $x^{(m_i+1)}(a_i) = A_i$, where i is one of the integers $1,\ldots,r_2$ have been obtained.

1.    Bebernes, J. W. and Jackson, L. K. "Infinite interval boundary value problems for $y'' = f(x,y)$", Duke Math. J. 34, 39-48 (1967).
2.    Belova, M. M. "Bounded solutions of non-linear differential equations of second order", Matematičeskii Sbornik 56, 469-503 (1962).

3.  Corduneanu, C. "Citeve probleme globale referitoare la ecuatiile diferentiale neliniare de ordinul al doilea", Academia Republicii Populare Romine. Filiala Iasi Studii si Cercetări Stiintifice. Matematică 7, 1-7 (1956).

4.  Corduneanu, C. "Existenta solutiilor marginite pentru unele ecuatii diferentiale de ordinul al doilea", Academia Republicii Populare Romine. Filiale Iasi Studii si Cercetări Stiintifice. Mathematică 8, 127-134 (1957).

5.  Gross, O. A. "The boundary value problem on an infinite interval", J. Math. Anal. Appl. 7. 100-109 (1963).

6.  Hartman, P. and Wintner, A. "On the non-increasing solutions of y" = f(x,y,y')", Amer J. Math. 73, 390-404 (1951).

7.  Jackson, L. K. "Subfunctions and second order ordinary differential inequalities", Advances in Math. 2, 307-363 (1968).

8.  Kneser, A. "Untersuchung und asymptotische Darstellung der Intergrale gewisser Differential gleichungen bei grossen Werthen des Arguments", Journal für die reine und angewandte Mathematik 116, 178-212 (1896).

9.  Mambriani, A. "Su un teoreme relativo alle equazioni differenziali ordinarie del 2° ordine" Atti della Accademia Nazionale dei Lincei. Classe di Scienze Fisiche, Matematiche e Naturali 9, 620-622 (1929).

10. Schuur, J. D. "The existence of proper solutions of a second order ordinary differential equation", Proc. Amer. Math. Soc. 17, 595-597 (1966).

11. Šeda, V. "On an application of the Stone theorem in the theory of differential equations", Časopis pro pěstováni matematiky, roč. 97, 183-189 (1972).

12. Umamaheswaram, S. "Boundary value problems for n-th order ordinary differential equations", Ph.D. Thesis, Univ. of Missouri, Columbia, 1973.

13. Umamaheswaram, S. "Boundary value problems for higher order differential equations", J. Diff. Equs. 18, 188-201 (1975).

14. Wong Pui-Kei "Existence and asymptotic behavior of proper solutions of a class of second order nonlinear differential equations", Pacific J. Math. 13, 737-760 (1963).

## 25. EQUATIONS WITH DEVIATING ARGUMENTS

In this final section, we shall first consider the following differential equation with deviating arguments

(25.1)  $$x^{(n)}(t) = f(t,<x>,<x'>,\ldots,<x^{(q)}>)$$

where $<x^{(i)}>$ stands for $(x^{(i)}(w_{i1}(t)),\ldots,x^{(i)}(w_{ip(i)}(t)))$. The function f is assumed to be continuous on $[a_1,a_r] \times R^N$, where $N = \sum\limits_{i=0}^{q} P(i)$. The functions $w_{ij}$; $1 \le j \le p(i)$, $0 \le i \le q$ are continuous on $[a_1,a_r]$. Let $\alpha = \min\{a_1, \min\limits_{a_1 \le t \le a_r} w_{ij}(t)$; $1 \le j \le p(i)$, $0 \le i \le q\}$ and $\beta = \max\{a_r,$

$\max\limits_{a_1 \le t \le a_r} w_{ij}(t)$; $1 \le j \le p(i)$, $0 \le i \le q\}$. With respect to the boundary conditions (2.4), we assume that the functions $\phi \in C^{(s_1)}[\alpha,a_1]$ and $\psi \in C^{(s_2)}[a_r,\beta]$, where $s_1 = \max\{k_1,q\}$ and $s_2 = \max\{k_r,q\}$ are given. If $\alpha = a_1$ and/or $\beta = a_r$, then we shall assume that the constants $\phi^{(i)}(a_1)$; $i = 0,1,\ldots,k_1$ and/or $\psi^{(i)}(a_r)$; $i = 0,1,\ldots,k_r$ are known. We seek a function $x \in B = C^{(n)}[\alpha,\beta] \cap C^{(s_1)}[\alpha,a_1] \cap C^{(n)}[a_1,a_r] \cap C^{(s_2)}[a_r,\beta]$, where $\eta = \min\{k_1,k_r\}$, which has the property that it satisfies the boundary conditions

$$x^{(j)}(t) = \phi^{(j)}(t); \ t \in [\alpha,a_1], \ 0 \le j \le s_1$$

(25.2)  $$x^{(j)}(a_i) = A_{j+1,i}; \ 2 \le i \le r-1, \ 0 \le j \le k_i$$

$$x^{(j)}(t) = \psi^{(j)}(t); \ t \in [a_r,\beta], \ 0 \le j \le s_2$$

and $x(t)$ is a solution of (25.1) in $[a_1,a_r]$.

With respect to the BVP (25.1), (25.2) we need the function

$\theta \in B$ which is defined as follows

$$\theta(t) = \begin{cases} \phi(t) & , \ t \in [\alpha, a_1] \\ \overline{P}_{n-1}(t), & t \in [a_1, a_r] \\ \psi(t) & , \ t \in [a_r, \beta] \end{cases}$$

where $\overline{P}_{n-1}(t)$ is same as $P_{n-1}(t)$ except that $A_{j+1,1} = \phi^{(j)}(a_1)$, $0 \le j \le k_1$ and $A_{j+1,r} = \psi^{(j)}(a_r)$, $0 \le j \le k_r$.

Theorem 25.1   Suppose that

(i)   $K_i > 0$, $0 \le i \le q$ are given real numbers and $Q$ is the maximum of $|f(t, x_{01}, \ldots, x_{0p(0)}, \ldots, x_{q1}, \ldots, x_{qp(q)})|$ on the compact set : $[a_1, a_r] \times D$, where

$$D = \{(x_{01}, \ldots, x_{qp(q)}) : |x_{i1}|, \ldots, |x_{ip(i)}| \le 2K_i, \ 0 \le i \le q\}$$

(ii)   $\max\limits_{\alpha \le t \le \beta} |\theta^{(i)}(t)| \le K_i$, $0 \le i \le q$ wherever $\theta^{(i)}(t)$ exists

(iii)   $(a_r - a_1) \le (K_i/QC_{n,i})^{1/n-i}$, $0 \le i \le q$.

Then, the BVP (25.1), (25.2) has a solution in $B$.

Proof.   Let $\nu = \min\{k_1, k_r, q\}$ and denote by $B_1$ the space

$$B_1 = C^{(\nu)}[\alpha, \beta] \cap C^{(q)}[\alpha, a_1] \cap C^{(q)}[a_1, a_r] \cap C^{(q)}[a_r, \beta].$$

We make $B_1$ into a Banach space by defining the norm of an element $x \in B_1$ by

$$\|x\| = \max\{ \max\limits_{\alpha \le t \le a_1} |x^{(i)}(t)|, \ \max\limits_{a_1 \le t \le a_r} |x^{(i)}(t)|, \ \max\limits_{a_r \le t \le \beta} |x^{(i)}(t)| ; \ 0 \le i \le q \}.$$

Now define a mapping $T$ on $B_1$ as follows : For each $x \in B_1$, let $Tx$ be the function

$$(25.3) \qquad (Tx)(t) = \theta(t) + \int_{a_1}^{a_r} \bar{g}_2(t,s) f(s,<x>,<x'>,\ldots,<x^{(q)}>) ds$$

where

$$\bar{g}_2(t,s) = \begin{cases} g_2(t,s), & t \in [a_1,a_r] \\ \\ 0, & t \notin [a_1,a_r]. \end{cases}$$

The following properties of T may easily be established :

(a) $(Tx)^{(j)}(t) = \theta^{(j)}(t) = \phi^{(j)}(t); \ t \in [\alpha,a_1], \ 0 \le j \le s_1.$

(b) $(Tx)^{(j)}(t) = \theta^{(j)}(t) = \psi^{(j)}(t); \ t \in [a_r,\beta], \ 0 \le j \le s_2.$

(c) $(Tx)(t)$ is n times continuously differentiable and

$(Tx)^{(n)}(t) = f(t,<x>,<x'>,\ldots,<x^{(q)}>)$ for all $t \in [a_1,a_r].$

(d) $T : \mathcal{B}_1 \to \mathcal{B}.$

(e) Fixed points of T are solutions of the BVP (25.1), (25.2).

(f) T is a completely continuous operator.

(g) $(Tx)(t) - \theta(t)$ satisfies conditions of Theorem 8.3 in $[a_1,a_r].$

Consider the closed convex subset S of the Banach space $\mathcal{B}_1$ defined by

$$S = \{x \in \mathcal{B}_1 : |x^{(i)}(t)| \le 2K_i; \ 0 \le i \le q \text{ wherever } x^{(i)}(t) \text{ exists}\}.$$

We shall show that T maps S into itself. For this, let $x \in S$. If $t \in [\alpha,a_1]$ or $t \in [a_r,\beta]$, then from the above properties (a) and (b) and hypothesis (ii), it is obvious that $|(Tx)^{(i)}(t)| \le K_i, \ 0 \le i \le q.$ If $t \in [a_1,a_r]$, then property (c) and hypothesis (i) provide $|(Tx)^{(n)}(t) - \theta^{(n)}(t)| = |(Tx)^{(n)}(t)| \le Q$, and hence from property (g) we can use Theorem 8.3 to obtain

$$|(Tx)^{(i)}(t) - \theta^{(i)}(t)| \le QC_{n,i}(a_r-a_1)^{n-i}$$

or

$$|(Tx)^{(i)}(t)| \le |\theta^{(i)}(t)| + QC_{n,i}(a_r-a_1)^{n-i}$$

which from hypothesis (iii) provides

$$\left| (Tx)^{(i)}(t) \right| \le K_i + K_i = 2K_i, \; 0 \le i \le q.$$

Thus, if $t \in [\alpha, \beta]$, we have $\left| (Tx)^{(i)} \right| \le 2K_i$, $0 \le i \le q$, i.e., T maps S into itself. It then follows from Schauder's fixed point theorem that T has a fixed point in S. From property (e) this fixed point is a solution of (25.1), (25.2) in $B$.

**Theorem 25.2**   Suppose that the function f satisfies the Lipschitz condition

$$(25.4) \qquad \left| f(t,<x>,<x'>,\ldots,<x^{(q)}>) - f(t,<y>,<y'>,\ldots,<y^{(q)}>) \right|$$

$$\le \sum_{i=0}^{q} \sum_{j=1}^{p(i)} L_{ij} \left| x^{(i)}(w_{ij}(t)) - y^{(i)}(w_{ij}(t)) \right|$$

for all $(t,<x>,\ldots,<x^q>)$, $(t,<y>,\ldots,<y^{(q)}>) \in [a_1,a_r] \times R^N$, and

$$(25.5) \qquad \tau = \left( \sum_{i=0}^{q} \sum_{j=1}^{p(i)} L_{ij} \right) \max_{0 \le i \le q} \{ C_{n,i}(a_r - a_1)^{n-i} \} < 1.$$

Then, the BVP (25.1), (25.2) has a unique solution in $B$.

**Proof.**   We shall show that the operator $T : B_1 \to B$ defined in (25.3) satisfies the conditions of Contraction Mapping Principle. For this, let $x,y \in B_1$, then we have

$$(Tx)(t) - (Ty)(t) = 0, \; t \in [\alpha, a_1] \text{ or } t \in [a_r, \beta]$$

$$= \int_{a_1}^{a_r} g_2(t,s)[f(s,<x>,\ldots,<x^{(q)}>) - f(s,<y>,\ldots,<y^{(q)}>)]ds,$$

$$t \in [a_1, a_r]$$

and hence, from Theorem 8.3 it follows that

$$|(Tx)^{(i)}(t) - (Ty)^{(i)}(t)|$$

$$\leq C_{n,i}(a_r-a_1)^{n-i} \times \max_{a_1 \leq t \leq a_r} |f(t,<x>,\ldots,<x^{(q)}>)-f(t,<y>,\ldots,<y^{(q)}>)|$$

$$\leq C_{n,i}(a_r-a_1)^{n-i} \left( \sum_{i=0}^{q} \sum_{j=1}^{p(i)} L_{ij} \right) \|x-y\| \; ; \; t \; \epsilon \; [a_1,a_r], \; 0 \leq i \leq q$$

which implies that

$$\|Tx - Ty\| \leq \tau \|x-y\|.$$

Since $\tau < 1$, the result follows.

Finally, for the differential equation (25.1) we note that the results similar to Theorems 25.1 and 25.2 for other BVPs can easily be stated.

In the rest of this section, we shall use shooting type method to prove the existence of a unique solution of the following BVP

$$(25.6) \qquad (\rho(t)x^{(n-1)}(t))' = f(t,x_t,\underline{x}(t))$$

$$x(t) = \phi(t), \; -r \leq t \leq 0$$

$$(25.7) \qquad x^{(i)}(0) = A_i \; ; \; i = 1,2,\ldots,n-2$$

$$x^{(p)}(T) = B, \; 0 \leq p \leq n-1 \text{ is fixed}$$

where $\phi \; \epsilon \; C_r = C([-r,0],R)$ which is endowed with the sup norm, and for any continuous function x defined on $[-r,T]$ and any $t \; \epsilon \; [0,T]$ the function $x_t$ is an element of $C_r$ defined by $x_t(s) = x(t+s)$, $-r \leq s \leq 0$, $t \; \epsilon \; [0,T]$. The function $\rho$ is positive and continuously differentiable on $[0,T]$. The function f is assumed to be continuous on $[0,T] \times C_r \times R^{q+1}$.

We shall need the following two lemmas.

Lemma 25.3   Let $x(t)$ and $y(t)$, $t \in [-r,T]$ be functions which have continuous $(n-1)$th order derivatives on $[0,T]$.  Assume further that

$$x(t) = y(t) = 0, \quad -r \le t \le 0$$

(25.8) $$x^{(i)}(0) = y^{(i)}(0) = 0, \quad 1 \le i \le n-2$$

$$x^{(n-1)}(0) = \gamma > \eta = y^{(n-1)}(0) > 0.$$

If $\varepsilon$ is any positive number such that $\frac{\gamma-\varepsilon}{\eta} > 1$ and $m(t)$ is defined by
$m(t) = x(t) - \frac{\gamma-\varepsilon}{\eta} y(t)$, then the following hold

   (i)    there exists a $\delta \in (0,T]$ such that for every $t \in (0,T]$,
          $m^{(i)}(t) > 0$, $0 \le i \le n-1$

   (ii)   if $m^{(n-1)}(t) > 0$ on $[0,t_0)$ and $m^{(n-1)}(t_0) = 0$, then

$$0 \le m(t) \le \frac{t^i}{i!} m^{(i)}(t), \quad 0 \le i \le n-2.$$

Proof.    Since $m^{(n-1)}(0) = \varepsilon > 0$, part (i) is obvious.  For $t \in [0,t_0)$,
Taylor's formula gives

$$m(t) = \frac{t^i}{i!} m^{(i)}(t*), \quad 0 < t* < t_0.$$

Part (ii) now follows from the increasing nature of $m^{(i)}(t)$ in $[0,t_0)$.

Lemma 25.4    Assume that

   (i)    $f(t,\phi,\underline{x})$ is a continuous function on $[0,T] \times C_r \times R^{q+1}$ and
          nondecreasing in each variable $\phi, x_0, x_1, \ldots, x_q$, also for
          $\lambda > 1$

(25.9) $$\lambda g(t,\phi,\underline{x}) \le g(t,\lambda\phi,\lambda\underline{x})$$

(ii)   $L(t,\phi)$ is a continuous function on $[0,T] \times C_r$ and for each $t \in [0,T]$, $L(t,\phi)$ is a linear bounded operator on $C_r$ into $R$, also there exists a continuous function $\ell(t)$ such that

$$|L(t,\phi)| \leq \ell(t) \|\phi\| ; \quad t \in [0,T], \ \phi \in C_r$$

(iii)  for every $(t,\phi,\underline{x}) \in [0,T] \times C_r \times R^{q+1}$ with $\phi \geq 0$, $x_i \geq 0$, $0 \leq i \leq q$

$$f(t,\phi,\underline{x}) \geq g(t,\phi,\underline{x}) + L(t,\phi) + \sum_{i=0}^{\tau} q_i(t)x^{(i)}(t)$$

where $\tau = \min\{q,n-2\}$ and $q_i(t) \geq 0$, $0 \leq i \leq \tau$ are continuous functions on $[0,T]$, also for every $t \in (0,T]$

(25.10)
$$\ell(t) - \sum_{i=0}^{\tau} q_i(t) \frac{i!}{t^i} < 0$$

(iv)   solutions $x(t,0,\gamma)$ and $y(t,0,\eta)$ of the differential equations (25.6) and

(25.11)
$$(\rho(t)y^{(n-1)})' = g(t,y_t,\underline{y}(t)) + L(t,y_t) + \sum_{i=0}^{\tau} q_i(t)y^{(i)}(t)$$

satisfying (25.8) respectively are assumed to exist on the whole interval $[0,T]$.

Then, for every $t \in [0,T]$

(25.12)     $x^{(i)}(t,0,\gamma) \geq \frac{\gamma}{\eta} y^{(i)}(t,0,\eta)$, $0 \leq i \leq n-1$.

Proof.   From Lemma 25.3 there exists a $\delta > 0$ such that for every $t \in (0,\delta]$, $m^{(i)}(t) > 0$, $0 \leq i \leq n-1$. We shall prove that $m^{(i)}(t) \geq 0$ on $[0,T]$. Let $t_0$ be the first positive zero of $m^{(n-1)}(t)$ on $[0,T]$, i.e.,

(25.13)     $m^{(n-1)}(t_0) = x^{(n-1)}(t_0,0,\gamma) - \frac{\gamma-\varepsilon}{\eta} y^{(n-1)}(t_0,0,\eta) = 0$

then, it is necessary that

$$(25.14) \qquad m^{(n)}(t_0) = x^{(n)}(t_0,0,\gamma) - \frac{\gamma-\varepsilon}{\eta} y^{(n)}(t_0,0,\eta) \leq 0.$$

Further, since $m(t)$ is increasing and positive function on $(0,t_0]$ and $m(t) = 0$ for $t \leq 0$, we find that

$$(25.15) \qquad \qquad \|m_{t_0}\| = m(t_0).$$

Now, from Lemma 25.3 and (25.13) - (25.15), we have the following:

$$f(t_0, x_{t_0}, \underline{x}(t_0))$$

$$= (\rho(t)x^{(n-1)}(t))'_{t=t_0}$$

$$= \rho'(t_0)x^{(n-1)}(t_0) + \rho(t_0)x^{(n)}(t_0) \leq \rho'(t_0)\frac{\gamma-\varepsilon}{\eta}y^{(n-1)}(t_0) + \rho(t_0)\frac{\gamma-\varepsilon}{\eta}y^{(n)}(t_0)$$

$$= \frac{\gamma-\varepsilon}{\eta}(\rho(t)y^{(n-1)}(t))'_{t=t_0}$$

$$= \frac{\gamma-\varepsilon}{\eta}[g(t_0,y_{t_0},\underline{y}(t_0)) + L(t_0,y_{t_0}) + \sum_{i=0}^{\tau} q_i(t_0)y^{(i)}(t_0)]$$

$$\leq g(t_0,\frac{\gamma-\varepsilon}{\eta}y_{t_0},\frac{\gamma-\varepsilon}{\eta}\underline{y}(t_0)) + L(t_0,\frac{\gamma-\varepsilon}{\eta}y_{t_0}) + \sum_{i=0}^{\tau} q_i(t_0)\frac{\gamma-\varepsilon}{\eta}y^{(i)}(t_0)$$

$$\leq g(t_0,x_{t_0},\underline{x}(t_0)) + L(t_0,\frac{\gamma-\varepsilon}{\eta}y_{t_0}) + \sum_{i=0}^{\tau} q_i(t_0)\frac{\gamma-\varepsilon}{\eta}y^{(i)}(t_0)$$

$$\leq f(t_0,x_{t_0},\underline{x}(t_0)) - L(t_0,x_{t_0}-\frac{\gamma-\varepsilon}{\eta}y_{t_0}) - \sum_{i=0}^{\tau} q_i(t_0)[x^{(i)}(t_0) - \frac{\gamma-\varepsilon}{\eta}y^{(i)}(t_0)]$$

$$\leq f(t_0,x_{t_0},\underline{x}(t_0)) + |L(t_0,m_{t_0})| - \sum_{i=0}^{\tau} q_i(t_0)m^{(i)}(t_0)$$

$$\leq f(t_0,x_{t_0},\underline{x}(t_0)) + \ell(t_0)\|m_{t_0}\| - \sum_{i=0}^{\tau} q_i(t_0)\frac{i!}{t_0^i}m(t_0)$$

$$\leq f(t_0,x_{t_0},\underline{x}(t_0)) + m(t_0)[\ell(t_0) - \sum_{i=0}^{\tau} q_i(t_0)\frac{i!}{t_0^i}]$$

which contradicts the condition (25.10), and hence (25.12) holds.

**Theorem 25.5**   Assume that

(i)   for all $(t,\phi,\underline{x})$, $(t,\psi,\underline{y})$ e $[0,T] \times C_r \times R^{q+1}$ with $\phi \geq \psi$,
$x_i \geq y_i$, $0 \leq i \leq q$

(25.16)   $f(t,\phi,\underline{x}) - f(t,\psi,\underline{y}) \geq g(t,\phi-\psi,\underline{x}-\underline{y}) + L(t,\phi-\psi) + \sum\limits_{i=0}^{\tau} q_i(t)(x_i-y_i)$

where $g, L$ and $q_i$, $0 \leq i \leq \tau$ satisfy the conditions of
Lemma 25.4

(ii)   for every $\eta > 0$, there exists a solution $y(t,0,\eta)$ of the
initial value problem (25.11), (25.8) on $[0,T]$ and
$y^{(p)}(t,0,\eta) > 0$ for every t e $[0,T]$

(iii)   for every $\gamma$ e R, $A = (A_1,\ldots,A_{n-2})$ e $R^{n-2}$ and $\phi$ e $C_r$ the
differential equation (25.6) satisfying

$$x(t) = \phi(t), \quad -r \leq t \leq 0$$

(25.17)
$$x^{(i)}(0) = A_i, \quad 1 \leq i \leq n-2$$

$$x^{(n-1)}(0) = \gamma$$

has a unique solution $x(t,\phi,A,\gamma)$ on $[0,T]$.

Then, there exists a unique solution of the BVP (25.6), (25.7).

**Proof.**   For $\Gamma$, $\gamma$ e R with $\dfrac{\Gamma-\gamma}{\eta} > 1$ consider the solutions $x(t,\phi,A,\Gamma)$ and
$x(t,\phi,A,\gamma)$ of the initial value problem (25.6), (25.17). We define,
$z(t,A,\Gamma,\gamma) = x(t, \phi, A, \Gamma) - x(t,\phi,A,\gamma)$, then $z(t,A,\Gamma,\gamma)$ is the solution
of the initial value problem

$$(\rho(t)z^{(n-1)}(t,A,\Gamma,\gamma))' = F(t,z_t(\cdot,A,\Gamma,\gamma),z(t,A,\Gamma,\gamma),\ldots,z^{(q)}(t,A,\Gamma,\gamma))$$

$$z(t,A,\Gamma,\gamma) = 0, \quad -r \leq t \leq 0$$

$$z^{(i)}(0,A,\Gamma,\gamma) = 0, \quad 1 \leq i \leq n-2$$

$$z^{(n-1)}(0,A,\Gamma,\gamma) = \Gamma - \gamma$$

where

$$F(t, z_t(\cdot, A, \Gamma, \gamma), z(t, A, \Gamma, \gamma), \ldots, z^{(q)}(t, A, \Gamma, \gamma))$$

$$= f(t, z_t(\cdot, A, \Gamma, \gamma) + x_t(\cdot, \phi, A, \gamma), z(t, A, \Gamma, \gamma) + x(t, \phi, A, \gamma),$$

$$\ldots, z^{(q)}(t, A, \Gamma, \gamma) + x^{(q)}(t, \phi, A, \gamma))$$

$$- f(t, x_t(\cdot, \phi, A, \gamma), x(t, \phi, A, \gamma), \ldots, x^{(q)}(t, \phi, A, \gamma)).$$

By (25.16) the function F satisfies the condition (iii) of Lemma 25.4. Thus, applying Lemma 25.4 for the solutions $z(t, A, \Gamma, \gamma)$ and $y(t, 0, \eta)$, we have $z^{(i)}(t, A, \Gamma, \gamma) \geq \frac{\Gamma - \gamma}{\eta} y^{(i)}(t, 0, \eta)$; $i = 0, 1, \ldots, n-1$ for every $t \in [0, T]$. Hence, in particular for $i = p$ and $t = T$, we get

$$z^{(p)}(T, A, \Gamma, \gamma) \geq \frac{\Gamma - \gamma}{\eta} y^{(p)}(T, 0, \eta) > 0$$

and consequently

$$x^{(p)}(T, \phi, A, \Gamma) - x^{(p)}(T, \phi, A, \gamma) > 0.$$

Thus, for fixed $\gamma \in R$

$$\lim_{\Gamma \to \infty} x^{(p)}(T, \phi, A, \Gamma) = \infty$$

and, for fixed $\Gamma \in R$

$$\lim_{\gamma \to -\infty} x^{(p)}(T, \phi, A, \gamma) = -\infty.$$

From the hypothesis (iii), $x^{(p)}(T, \phi, A, \gamma) - B$ is a continuous function of $\gamma$ (see [9]), and its range must be the whole real line R. Hence, there exists a $\gamma_p \in R$ such that $x^{(p)}(T, \phi, A, \gamma_p) = B$. This $x(t, \phi, A, \gamma_p)$ is a solution of the BVP (25.6), (25.7).

Next, let $x_1(t)$ and $x_2(t)$ be two solutions of the BVP (25.6),(25.7) then from the hypothesis (iii) it is necessary that $x_1^{(n-1)}(0) \neq x_2^{(n-1)}(0)$. Let $x_1^{(n-1)}(0) = \Gamma > \gamma = x_2^{(n-1)}(0)$, then as in the existence proof we easily arrive at the inequality $x_1^{(p)}(T) - x_2^{(p)}(T) > 0$. This contradiction completes the proof of our theorem.

## COMMENTS AND BIBLIOGRAPHY

Theorems 25.1 and 25.2 are taken from [3] and cover some of the results for second and third order delay differential equations given in [1,2,5,6,8,10-13]. Theorem 25.5 generalizes and improves the results obtained in [7,14-16]. Some similar results for the discrete BVPs are available in [4].

1. Agarwal, R. P. "Boundary value problems for differential equations with deviating arguments", J. Math. Phyl. Sci. 6, 425-438 (1972).
2. Agarwal, R. P. "Existence and uniqueness for nonlinear functional differential equations", Indian J. Pure and Appl. Math. 7, 933-938 (1976).
3. Agarwal, R. P. "Boundary value problems for differential equations with deviating arguments", Bull. Inst. Math. Acad. Sinica 9, 63-67 (1981).
4. Agarwal, R. P. "Initial and boundary value problems for n-th order difference equations", Mathematica Slovaca, to appear.
5. Aliev, R. M. "On a boundary value problem for second order linear differential equations with retarded argument (Russian)", in Proc. Second Scientific Conference at Moscow's People's Friendship University, 15-16, 1966.
6. Bernfeld, S., Ladde, G. and Lakshmikantham, V. "Nonlinear boundary value problems and several Lyapunov functions for functional differential equations", Boll. U.M.I. 10, 602-613 (1974).
7. De Nevers, K. and Schmitt, K. "An application of the shooting method to boundary value problems for second order delay equations", J. Math. Anal. Appl. 36, 588-597 (1971).
8. Grimm, L. J. and Schmitt, K. "Boundary value problems for differential equations with deviating arguments", Aequationes Math. 3, 24-38 (1969).
9. Hale, J. K. Functional Differential Equations, Springer-Verlag, New York, 1971.
10. Kalinowski, J. "On the convergence of an iterative sequence to the solution of a system of ordinary differential equations with deviating arguments", Demonstratio Mathematica 9, 77-93 (1976).

11. Kalinowski, J. "Two-point boundary problems for some systems of ordinary differential equations of second order with deviating arguments (Russian)", Ann. Polon. Math. 30, 71-76 (1974).

12. Kamenskii, G. A. "Boundary value problems for nonlinear equations with perturbed arguments (Russian)", Nauk-Dokl. Vyssh. Shkoly Fiz. Mat. Nauki 2, 60-66 (1958).

13. Kamenskii, G. A. "Boundary value problems for nonlinear differential equations with deviating arguments of neutral type (Russian)", Trudy Sem. Teor. Diff. Urav. Otklon. Arg. 1, 47-51 (1962).

14. Ohkohchi, S. "A boundary value problem for delay differential equations", Hiroshima Math. J. 7, 379-385 (1977).

15. Schmitt, K. "On solutions of nonlinear differential equations with deviating arguments", SIAM J. Appl. Math. 17, 1171-1176 (1969).

16. Sficas, Y. G. and Ntouyas, S. K. "A boundary value problem for n-th order functional differential equations", Nonlinear Analysis: Theory, Methods and Appl. 5, 325-335 (1981).

# NAME INDEX

306